Electrical installation
and workshop technology

Volume one Third Edition Metric

Electrical installation and workshop technology

Volume one Third Edition Metric

F. G. Thompson,
T. Eng (CEI), FITE, MASEE

Department of Engineering
Lews Castle College, Stornoway

Illustrative material by

J. H. Smith, AMITE

Late Department of Electrical & Electronic Engineering,
Inverness Technical College, Inverness

Longman
London and New York

LONGMAN GROUP LIMITED
London and New York

Associated companies, branches and representatives
throughout the world

© Longman Group Limited 1968, 1973, 1978

First published 1968
Second edition Metric 1973
Second impression 1974
Third edition 1978

ISBN 0 582 41609 4

Library of Congress Cataloging in Publication Data

Thompson, Francis G.
    Electrical installation and workshop technology.

    Includes index.
    1. Electric wiring – Handbooks, manuals, etc.
I. Title.
TK3201.T47   1978   621.319'24   77-21748
ISBN 0 582 41609 4

Printed in Great Britain by
Richard Clay (The Chaucer Press) Ltd, Bungay, Suffolk

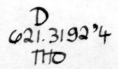

Contents

Part B   Electrical workshop technology

Part C   Appendices

List of illustrations

Preface

This textbook is intended mainly for students who are studying the subject of Electrical Installation Work, contained within the City and Guilds of London Institute Course 236, leading to the award of the Part I and Part II Certificates in the subject. The engineering content of the book is aimed at those whose 'on-the-job' training lacks a significant workshop element. The book will also be found useful to those students who are taking the preliminary stages of the C & G Technicians' Course, 280/281, Course 231/232, Electrical & Electronic Craft Studies, and Course 200 in Basic Engineering Craft Studies, Part I.

The material presented herein covers the two-years' requirements of the Course 236, Part I. It also serves as a good rounded introduction to the year's work which leads to Part II, of CGLI Course 236, leading to the award of the Electrician's Certificate. Emphasis has been placed on the practical aspect of electrical science theory, to give the fully-comprehensive background knowledge of installation technology and engineering which is necessary if a student is to become at all competent in his chosen career field.

Opportunity has been taken to 'liberalise' the electrical installation technology material by presenting certain aspects which usually come into the category of related studies. This has been done so that a broad-based understanding can be obtained by the student of his work, and the implications it has in socio-economic fields of human activity.

Though the book is specifically aimed at the students of the courses mentioned, the material will be found useful as an aid to teachers in technical colleges—and to those teachers in secondary schools faced with the problem of introducing vocational aspects of an electrical nature into the 'craft-orientated' courses. To increase the value of the book to teachers, material has been included for the effective consolidation of learning.

The emphasis throughout the book is on the subject of 'installation engineering', which includes all the work which the electrician in the highly-automated society of today is called on to do. The new, up-and-coming breed of installation engineers are basically electricians with a specialist interest in, and knowledge of, all forms of current-using apparatus and the systems used to take electrical supplies to these equipments. Installation engineering is already a recognised branch of electrical engineering activity. The electrician of today and tomorrow will find himself to be a man responsible for the application of a specialist knowledge to the field of electricity utilisation. The work of the electrician is now being defined within the electrical engineering industry. It is hoped that this book will go some useful way in catering for his early technical-education needs for a long time to come.

In this new edition opportunity has been taken to strengthen the text and to introduce those aspects of metrication which most directly affect the work of the student or apprentice in electrical installation work. The author wishes to thank readers who have taken the trouble to write to him and offer comments and suggestions for the new edition, which are now incorporated. The sales history of the book has indicated that it is catering for a demand which it is hoped the new edition will continue to satisfy.

Finally, the author wishes to record his sincere thanks to those who helped in some way the progress of this book to its fruition; particularly his colleague Mr A. G. MacDonald who read the manuscript and made many helpful suggestions. Also, thanks are due to the following who, at some trouble, time and expense to themselves, went out of their way to provide illustrative and other matter for the book:

City and Guilds of London Institute
Institution of Electrical Engineers
Association of Supervising Electrical Engineers
British Standards Institution
British Insulated Callender's Cables Limited
Davis Sheet-metal Engineering Co. Limited
Messrs Ecko-Ensign Limited
J. A. Crabtree & Co. Limited
Walsall Conduits Limited
Evershed & Vignoles Limited
Messrs Klockner-Möeller

Part A

Electrical installation technology

1

The career of the electrician

The spectacular increase in the use of electrical energy for virtually all domestic and industrial purposes over the last half century or so is proof enough that the electrical industry plays a most prominent part in the economy of the country. The range of careers which the industry offers is extremely wide. But whatever the career, a fundamental knowledge of electrical engineering, its science and its technology, is necessary for any progress in the career.

The electrician of today plays perhaps one of the most important roles within the electrical industry, and not only in the matter of providing a labour force of skills and abilities on many levels. He is most definitely a key man with a fair degree of responsibility for work which can be carried out satisfactorily only with a background of sound technical knowledge.

Whatever the particular field of employment—supply, manufacturing or contracting—possession of technical knowledge forms the basis for the performance of the many varied tasks which today's electrician is called on by industry and the householder to do—and do well.

Today's electrician works in the supply industry to provide services associated with the generation, transmission and distribution of electrical energy. The 'link-man' between the supply industry and the user of electricity is the electrician employed in the vast field of employment known as the contracting industry. The function of electric wiring is basically to carry electrical energy to a point of use where it is converted into some other form of energy: into light, heat or mechanical power. Wiring, and so the contracting electrician, thus occupies the unique position of being an essential link between the supply authorities on the one hand and the makers and users of all kinds of appliances and apparatus on the other. Once, in years gone by, the idea existed that 'wiring' was hardly a respectable occupation for any but those without technical qualifications. The image has changed, however, and the electrician of today is required to have a minimum certificate to indicate his ability to do his job to a certain standard. Indeed, in some countries overseas, it is impossible for anyone to set up a career as an electrician unless he has passed a theoretical and practical examination: this is a legal requirement.

Electricity in this country, indeed in any country, is much more than a national asset. It has a deep social importance. It has been the main influence for good in the field of improvement in our standards of living. Electricity has proved to be the most flexible form of power in existence: it can be generated easily and transmitted to whatever or whomever requires it, and in whatever quantity it may be required. The historian Trevelyan has said that the social scene grows out of economic conditions. It is true to say that with the aid of electricity, a livelihood and a way of life without drudgery has been won, with an attendant enrichment of the social scene. That electricity is now commonplace in our lives is proved by the fact that the ordinary user regards it as essential to his daily life and living, and his work, and accepts it completely without question.

Offering, as it does to the user, a clean and effortless way of life, electricity also offers to the electrician a means whereby an interesting and satisfying living can be made. The opportunities available are legion, but particularly to those who qualify for promotion by study. Emphases nowadays placed on the possession of some minimum qualifications are so insistent that progress in any electrical career is virtually impossible without it. But theory alone is not enough. Practical experience is also a vital necessity; for not only must a job be done, but it must be understood thoroughly how and why it is done.

So far as the electrician of today is concerned, the bulk of the practical training is through the medium of an apprenticeship lasting four years. This period is important for any boy. For it offers not only a means of earning some money while training for a career, but is very much like a foundation-stone for future progress in that career. Faulty instruction, for instance, may mar the chances which come up in later life, and may indeed leave a recognisable mark—that of the incompetent tradesman. Attention to all approved instruction received during the apprenticeship period will therefore ensure that the career progresses surely and steadily to attain whatever positions are offered within the industry.

Technical knowledge and experience are thus essentials which form a background to responsibility. In the contracting industry, the delegation of responsibility is a necessary feature of any class of work undertaken. On a building site, for instance, one can see the various grades of responsibility handed over to qualified electricians: chargehand, foreman, site engineer and so on. Responsibility is not only for the quality of performance of the men being supervised, but also for the quality of the materials used in the contract and how they are applied.

In the manufacturing industry, responsibility is given to those who have proved themselves able to ensure, by applying their knowledge and experience, that electrical energy is used in the most efficient way, so increasing productivity in the application and maintenance aspects.

The basic requirement for any worthwhile career in the electrical industry is the possession of some qualification. So far as the electrician is concerned, the present-day accepted qualifications are those generally provided by the City and Guilds of London Institute: the Certificates in Electrical Installation Work, the Electrical Technicians Certificate, the Certificate for Electrical Craft Practice and the Electrical Installation Technicians Certificate.

Possession of any of these qualifications indicates at least a minimum standard of technical skill and ability to do a job satisfactorily.

Over the years the Electrical Installation Course offered by the City and Guilds of London Institute has undergone a number of basic changes which are intended to strengthen the quality of entrants into the trade and out of it when the apprenticeship has been completed. These changes have been made by training boards, colleges and examination bodies. The new Electrical Installation Work Course, No 236 (old 235) replaces the former 'A' and 'B' Courses. The New Course leads to Part II in, notionally, three years of part-time day-release (or its equivalent in block release over two or three years). Part I is offered on an examination basis and is sat en route, usually about two-thirds of the way through the Part II Course. For some candidates (depending on their previous attainment, their potential progress rates and their concurrent training opportunities) a somewhat shorter or longer period might be appropriate. In some cases, Part I will take the whole of three years of day release. The Certificates awarded by the City and Guilds Institute are part of the grading requirements of the Joint Industry Boards.

The Part I examination follows a course in which the emphasis is on practical training and knowledge to enable the possessor of the Certificate, the practical electrician, to carry out his duties efficiently under supervision. The Course supplements industrial experience and training which the student is given as an apprentice.

The Part II award is for the Electrician's Certificate and recognition

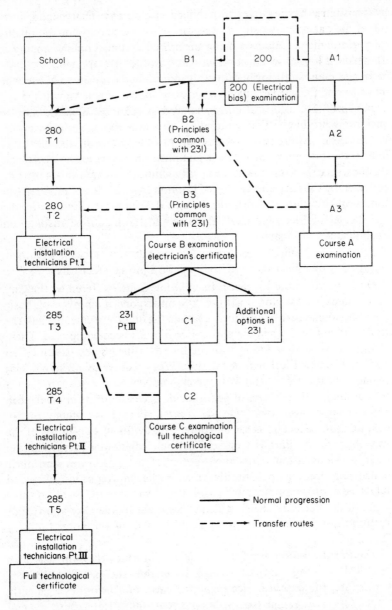

Figure 1.1. Diagram of route to become an Electrician (Craft Practice Certificate), an Approved Electrician (Electricians' Certificate), and a Technician Electrician (Full Technological Certificate). (*By courtesy of the City and Guilds of London Institute.*)

by the electrical contracting industry as an 'approved' electrician. This status indicates the ability to operate as a competent and qualified electrician without detailed supervision. This is the minimum standard to which all electrical installation apprentices should aspire, for it becomes a real step in the student's progression to positions of limited responsibility.

The Course 'C' is intended for those who aim for positions of definite responsibility. The Course is designed to afford the requisite technical knowledge for a 'technician electrician' (as defined by the Joint Industry Board), who will be required to plan and supervise general types of electrical installation work. Other CGLI Courses offer opportunities for able electrical students to specialise in planning and estimating for large contracting firms.

2

Historical review of installation work

As one might expect to find in the early beginnings of any industry, the application, and the methods of application, of electricity for lighting, heating and motive power was primitive in the extreme. Large-scale application of electrical energy was slow to develop. The first wide use of it was for lighting in houses, shops and offices. By the 1870s, electric lighting had advanced from being a curiosity to something with a definite practical future. Arc lamps were the first form of lighting, particularly for the illumination of main streets. When the incandescent-filament lamp appeared on the scene electric lighting took on such a prominence that it severely threatened the use of gas for this purpose. But it was not until cheap and reliable metal-filament lamps were produced that electric lighting found a place in every home in the land. Even then, because of the low power of these early filament lamps, shop windows continued for some time to be lighted externally by arc lamps suspended from the fronts of buildings.

The earliest application of electrical energy as an agent for motive power in industry is still electricity's greatest contribution to industrial expansion. The year 1900 has been regarded as a time when industrialists awakened to the potential of the new form of power.

Electricity was first used in mining for pumping. In the iron and steel industry, by 1917, electric furnaces of both the arc and induction type were producing over 100,000 tons of ingots and castings. The first all-welded ship was constructed in 1920; and the other ship-building processes were operated by electric motor power for punching, shearing, drilling machines and woodworking machinery.

The first electric motor drives in light industries were in the form of one motor-unit per line of shafting. Each motor was started once a day and continued to run throughout the whole working day in one direction at a constant speed. All the various machines driven from the shafting were started, stopped, reversed or changed in direction and speed by mechanical means. The development of integral electric drives, with provisions for starting, stopping and speed changes, led to the extensive use of the motor in small kilowatt ranges to drive an associated single machine, e.g. a lathe. One of the pioneers in the use of motors was the firm of Bruce Peebles, Edinburgh. The firm supplied in the 1890s, a number of weatherproof, totally-enclosed motors for quarries in Dumfriesshire, believed to be among the first of their type in Britain. The first electric winder ever built in Britain was supplied in 1905 to a Lanark oil concern. Railway electrification started as long ago as 1883, but it was not until long after the turn of this century that any major development took place.

Electrical installations in the early days were quite primitive and often dangerous. It is on record that in 1881, the installation in Hatfield House, London, was carried out by an aristocratic amateur. That the installation was dangerous did not perturb visitors to the house who '. . . when the naked wires on the gallery ceiling broke into flame . . . nonchalantly threw up cushions to put out the fire and then went on with their conversation'.

Many names of the early electrical pioneers survive today. Julius Sax began to make electric bells in 1855, and later supplied the telephone with which Queen Victoria spoke between Osborne, in the Isle of Wight, and Southampton in 1878. He founded one of the earliest purely electrical manufacturing firms which exists today and still makes bells and signalling equipment.

The General Electric Company had its origins in the 1880s, as a Company which was able to supply every single item which went to form a complete electrical installation. In addition it was guaranteed that all the components offered for sale were technically suited to each other, were of adequate quality and were offered at an economic price.

Specialising in lighting, Falk Stadelmann & Co. Ltd began by marketing improved designs of oil lamps, then gas fittings, and ultimately electric lighting fittings.

Cable makers W. T. Glover & Co. were pioneers in the wire field. Glover was originally a designer of textile machinery, but by 1868 he was also making braided steel wires for the then fashionable crinolines. From this type of wire it was a natural step to the production of insulated conductors for electrical purposes. At the Crystal Palace Exhibition in 1885 he showed a great range of cables; he was also responsible for the wiring of the exhibition.

The well-known J. & P. firm (Johnson & Phillips) began with making telegraphic equipment, extended to generators and arc lamps, and then to power supply.

The coverings for the insulation of wires in the early days included textiles and gutta-percha. Progress in insulation provisions for cables was made when vulcanised-rubber was introduced, and it is still used today. The first application of a lead sheath to rubber-insulated cables was made by Siemens Brothers. The manner in which we name cables was also a product of Siemens, whose early system was to give a cable a certain length related to a standard resistance of 0·1 ohm. Thus a No. 90 cable in his catalogue was a cable of which 90 yards had a resistance of 0·1 ohm. Cable sizes were also generally known by the Standard Wire Gauge.

For many years ordinary VRI cables made up about 95 per cent of all installations. They were used first in wood casing, and then in conduit. Wood casing was a very early invention. It was introduced to separate conductors, this separation being considered a necessary safeguard against the two wires touching and so causing fire. Choosing a cable at the turn of the century was quite a task. From one catalogue alone, one could choose from fifty-eight sizes of wire, with no less than fourteen different grades of rubber insulation. The grades were described by such terms as light, high, medium or best insulation. Nowadays there are two grades of insulation: up to 600V and 600V/1,000V. And the sizes of cables have been reduced to a more practicable twenty-two.

During the 1890s the practice of using paper as an insulating material for cables was well established. One of the earliest makers was the company which later became a member of the present-day BICC Group. The idea of using paper as an insulation material came from America to Britain where it formed part of the first wiring system for domestic premises. This was twin lead-sheathed cables. Bases for switches and other accessories associated with the system were of cast solder, to which the cable sheathing was wiped, and then all joints sealed with a compound. The compound was necessary because the paper insulation when dry tends to absorb moisture.

In 1911, the famous 'Henley Wiring System' came on the market. It comprised flat-twin cables with a lead-alloy sheath. Special junction boxes, if properly fixed, automatically effected good electrical continuity. The insulation was rubber. It became very popular. Indeed, it proved so easy to instal that a lot of unqualified people appeared on the contracting scene as 'electricians'. When it received the approval of the IEE Rules, it became an established wiring system and is still in use today.

At the time the lead-sheathed system made its first appearance, another rival wiring system also came onto the scene. This was the CTS system (cab-tyre sheathed). It arose out of the idea that if a rubber product could be used to stand up to the wear and tear of motor-car tyres on roads, then the material could well be applied to cover cables. The CTS name eventually gave way to TRS (tough-rubber sheath), when the rubber-sheathed cable system came into general use.

The main competitor to rubber as an insulating material appeared in the late 1930s. This material was PVC (polyvinylchloride), a synthetic material which came from Germany. The material, though inferior to rubber so far as elastic properties were concerned, could withstand the effects of both oil and sunlight. During the Second World War PVC, used both as wire insulation and the protective sheath, became well established.

As experience increased with the use of TRS cables, it was made the basis of modified wiring systems. The first of these was the Callender Farm-Wiring System introduced in 1937. This was Tough-Rubber sheathed cables with a semi-embedded braiding treated with a green-coloured compound. This system combined the properties of ordinary TRS and HSOS (House-service overhead system) cables.

So far as conductor material was concerned, copper was the most widely used. But aluminium was also applied as a conductor material. Aluminium, which has excellent electrical properties, has been pro-duced on a large commercial scale since about 1890. Overhead lines of aluminium were first installed in 1898. Rubber-insulated aluminium cables of $3/\cdot036''$ and $3/\cdot045''$ were made to the order of the British Aluminium Company and used in the early years of this century for the wiring of the staff quarters at Kinlochleven in Argyllshire. Despite the fact that lead and lead-alloy proved to be of great value in the sheathing of cables, aluminium was looked to for a sheath of, in particular, light weight. Many experiments were carried out before a reliable system of aluminium-sheathed cable could be put on the market.

Perhaps one of the most interesting systems of wiring to come into existence was the MICS (mineral-insulated copper-sheathed cable) which used compressed magnesium oxide as the insulation, and had a copper sheath and copper conductors. The cable was first developed in 1897 and was first produced in France. It has been made in Britain since 1937, firstly by Pyrotenax Ltd, and later by other firms. Mineral insula-tion has also been used with conductors and sheathing of aluminium.

One of the first suggestions for steel used for conduit was made in 1883. It was then called 'small iron tubes'. However, the first conduits were of bitumised paper. Steel for conduits did not appear on the wiring scene until about 1895. The revolution in conduit wiring dates from 1897, and is associated with the name 'Simplex' which is common enough today. It is said that the inventor, L. M. Waterhouse, got the idea of close-joint conduit by spending a sleepless night in a hotel bedroom staring at the bottom rail of his iron bedstead. In 1898 he began the production of light gauge close-joint conduits. A year later the screwed-conduit system was introduced.

Non-ferrous conduits were also a feature of the wiring scene. Heavy-gauge copper tubes were used for the wiring of the Rylands Library in Manchester in 1886. Aluminium conduit, though suggested during the 1920s, did not appear on the market until steel became a valuable material for munitions during the Second World War.

Insulated conduits also were used for many applications in installation work, and are still used to meet some particular installation conditions. The 'Gilflex' system, for instance, makes use of a PVC tube which can be bent cold, compared with earlier material which required the use of heat for bending.

Accessories for use with wiring systems were the subject of many experiments; many interesting designs came onto the market for the electrician to use in his work. When lighting became popular, there arose a need for the individual control of each lamp from its own control point. The 'branch switch' was used for this purpose. The term 'switch' came over to this country from America, from railway terms which indicated a railway 'point', where a train could be 'switched' from one set of tracks to another. The 'switch', so far as the electric circuit was concerned, thus came to mean a device which could switch an electric current from one circuit to another.

It was Thomas Edison who, in addition to pioneering the incandescent lamp, gave much thought to the provision of branch switches in circuit wiring. The term 'branch' meant a tee-off from a main cable to feed small current-using items. The earliest switches were of the 'turn' type, in which the contacts were wiped together in a rotary motion to make the circuit. The first switches were really crude efforts: made of wood and with no positive ON or OFF position. Indeed, it was usual practice to make an inefficient contact to produce an arc to 'dim' the lights! Needless to say, this misuse of the early switches, in conjunction with their wooden construction, led to many fires. But new materials were brought forward for switch construction such as slate, marble, and later, porcelain. Movements were also made more positive with definite ON and OFF positions.

The 'turn' switch eventually gave way to the 'tumbler' switch in popularity. It came into regular use about 1890. Where the name 'tumbler' originated is not clear; there are many sources, including the similarity of the switch action to the antics of Tumbler Pigeons. Many accessory names which are household words to the electricians of today appeared at the turn of the century: Verity's, McGeoch, Tucker and Crabtree. Further developments to produce the semi-recessed, the flush, the ac only, and the 'silent' switch proceeded apace. The switches of today are indeed of long and worthy pedigrees.

It was one thing to produce a lamp operated from electricity. It was quite another thing to devise a way in which the lamp could be held securely while current was flowing in its circuit. The first lamps were fitted with wire tails for joining to terminal screws. It was Thomas Edison who introduced, in 1880, the screw-cap which still bears his name. It is said he got the idea from the stoppers fitted to kerosene cans of the time. Like many another really good idea, it superseded all its competitive lampholders and its use extended through America and Europe. In Britain, however, it was not popular. The bayonet-cap type of lamp-holder was introduced by the Edison & Swan Co. about 1886. The early type was soon improved to the lampholders we know today.

Ceiling roses, too, have an interesting history; some of the first types incorporated fuses. The first rose for direct attachment to conduit came out in the early 1900s, introduced by Dorman & Smith Ltd.

The first patent for a plug-and-socket was brought out by Lord Kelvin, a pioneer of electric wiring systems and wiring accessories. The accessory was used mainly for lamp loads at first, and so carried very small currents. However, domestic appliances were beginning to appear on the market, which meant that sockets had to carry heavier currents. Two popular items were irons and curling-tong heaters. Shuttered sockets were designed by Crompton in 1893. The modern shuttered type of socket appeared as a prototype in 1905, introduced by 'Diamond H'. Many sockets were individually fused, a practice which was later extended to the provision of a fuse in the plug. These fuses were, however, only a small piece of wire between two terminals and caused such a lot of trouble that in 1911 the Institution of Electrical Engineers banned their use. One firm which came into existence with the socket-and-plug was M.K. Electric Ltd. The initials were for 'Multi-Kontakt' and associated with a type of socket-outlet which eventually became the standard design for this accessory. It was Scholes, under the name of 'Wylex', who introduced a revolutionary design of plug-and-socket: a hollow circular earth pin and rectangular current-carrying pins. This was really the first attempt to 'polarise', or to differentiate between live, earth and neutral pins.

One of the earliest accessories to have a cartridge fuse incorporated in it was the plug produced by Dorman & Smith Ltd. The fuse actually formed one of the pins, and could be screwed in or out when replacement was necessary. It is a rather long cry from those pioneering days to the present system of standard socket-outlets and plugs.

Early fuses consisted of lead wires, lead being used because of its low melting point. Generally, devices which contained fuses were called 'cutouts', a term still used today for the item in the sequence of supply-control equipment entering a building. Once the idea caught on of providing protection for a circuit in the form of fuses, brains went to work to design fuses and fusegear. Control gear first appeared encased in wood. But ironclad versions made their due appearance, particularly for industial use during the nineties. They were usually called 'motor switches', and had their blades and contacts mounted on a slate panel. Among the first companies in the switchgear field were Bill & Co., Sanders & Co. and the MEM Co., whose 'Kantark' fuses are so well known today. In 1928 this Company introduced the 'splitter' which effected a useful economy in many of the smaller installations.

It was not until the 1930s that the distribution of electricity in buildings by means of busbars came into fashion, though the system had been used as far back as about 1880, particularly for street mains. In 1935 the English Electric Co. introduced a busbar trunking system designed to meet the needs of the motor-car industry. It provided the overhead distribution of electricity into which system individual machines could be tapped wherever required; this idea caught on and designs were produced and put onto the market by Marryat & Place, GEC and Ottermill.

Trunking came into fashion mainly because the larger sizes of conduit proved to be expensive and troublesome to instal. One of the first trunking types to be produced was the 'spring conduit' of the Manchester firm of Key Engineering. They showed it for the first time at an electrical exhibition in 1908. It was semi-circular steel troughing with edges formed in such a way that they remained quite secure by a spring action after being pressed into contact. But it was not until about 1930 that the idea took root and is now established as a standard wiring system.

The story of electric wiring, its systems and accessories tells an important aspect in the history of industrial development and in the

history of social progress. The inventiveness of the old electrical person-
alities, Crompton, Swan, Edison, Kelvin and many others is well worth
noting; for it is from their brain-children that the present-day electrical
contracting industry has evolved to become one of the most important
sections of activity in electrical engineering. For those who are interested
in details of the evolution and development of electric wiring systems
and accessories, good reading can be found in the book by J. Mellanby:
The History of Electric Wiring (MacDonald, London).

3

Historical review of
wiring regulations

The history of the development of non-legal and statutory rules and regulations for the wiring of buildings is no less interesting than that of wiring systems and accessories. When electrical energy received a utilisation impetus from the invention of the incandescent lamp, many set themselves up as electricians or electrical wiremen.* Others were gas plumbers who indulged in the installation of electrics as a matter of normal course. This was all very well: the contracting industry had to get started in some way, however ragged. But with so many amateurs troubles were bound to multiply. And they did. It was not long before arc lamps, sparking commutators, and badly-insulated conductors contributed to fires. It was the insurance companies which gave their attention to the fire risk inherent in the electrical installations of the 1880s. Foremost among these was the Phoenix Assurance Co., whose engineer, Mr Heaphy, was told to investigate the situation and draw up a report on his findings.

The result was the Phoenix Rules of 1882. These Rules were produced just a few months after those of the American Board of Fire Underwriters who are credited with the issue of the first wiring rules in the world.

The Phoenix Rules were, however, the better set and went through many editions before revision was thought necessary. That these Rules contributed to a better standard of wiring, and introduced a high factor of safety in the electrical wiring and equipment of buildings, was indicated by a report in 1892 which showed the high incidence of electrical fires in the USA and the comparative freedom from fires of electrical origin in Britain.

Three months after the issue of the Phoenix Rules for wiring in 1882, the Society of Telegraph Engineers and Electricians (now the Institution of Electrical Engineers) issued the first edition of *Rules and Regulations*

---

* The first wiremen were originally carpenters or bell-hangers, who instructed in the 'new system of electric lighting'.

for the Prevention of Fire Risks arising from Electric Lighting. These rules were drawn up by a committee of eighteen men which included some of the famous names of the day: Lord Kelvin, Siemens and Crompton. The Rules, however, were subjected to some criticism. Compared with the Phoenix Rules they left much to be desired. But the Society was working on the basis of laying down a set of principles rather than, as Heaphy did, drawing up a guide or 'Code of Practice'. A second edition of the Society's Rules was issued in 1888. The third edition was issued in 1897 and entitled *General Rules recommended for Wiring for the Supply of Electrical Energy.*

The Rules have since been revised at fairly regular intervals as new developments and the results of experience can be written in for the considered attention of all those concerned with the electrical equipment of buildings. Basically the regulations were intended to act as a guide for electricians and others to provide a degree of safety in the use of electricity by inexperienced persons such as householders. The regulations were, and still are, not legal; that is, they cannot be enforced by the law of the land. Despite this apparent loophole, the regulations are accepted as a guide to the practice of installation work which will ensure, at the very least, a minimum standard of work. The Institution of Electrical Engineers (IEE) was not alone in the insistence of good standards in electrical installation work. In 1905, the Electrical Trades Union, through the London District Committee, in a letter to the Phoenix Assurance Co., said '. . . they view with alarm the large extent to which bad work is now being carried out by electric light contractors. . . . As the carrying out of bad work is attended by fires and other risks, besides injuring the Trade, they respectfully ask you to . . . uphold a higher standard of work'.

The legislation embodied in the Factory and Workshop Acts of 1901 and 1907 had a considerable influence on wiring practice. In the latter Act it was recognised for the first time that the generation, distribution and use of electricity in industrial premises could be dangerous. To control electricity in factories and other premises a draft set of Regulations was later to be incorporated into Statutory requirements.

While the IEE and the Statutory regulations were making their positions stronger, the British Standards Institution brought out, and is still issuing, Codes of Practice to provide what is regarded as guidance to good practice. The position of the Statutory Regulations in this country is that they form the primary requirements which must by law be satisfied. The IEE Regulations and Codes of Practice indicate supplementary requirements. However, it is accepted that if an installation is carried out in accordance with the IEE Wiring Regulations, then it generally fulfils the requirements of the Electricity Supply Regulations. This means that a supply authority can insist upon all electrical work to be carried out to the standard of the IEE Regulations, but cannot insist on a standard which is in excess of the IEE requirements.

The position of the IEE 'Regs' as they are popularly called, is that of being the installation engineer's 'bible'. Because the Regulations cover the whole field of installation work, and if they are complied with, it is certain that the resultant electrical installation will meet the requirements of all interested parties. There are, however, certain types of electrical installations which require special attention to prevent fires and accidents. These include mines, cinemas, theatres, factories and places where there are exceptional risks.

The following list gives the principal regulations which cover electricity supply and electrical installations:

Non-Statutory Regulations:

1. Institution of Electrical Engineers Regulations for the Electrical Equipment of Buildings—This covers industrial and domestic electrical installation work in buildings.
2. The Institute of Petroleum Electrical Code, 1963—This indicates special safety requirements in the petroleum industry, including protection from lightning and static. It is supplementary to the IEE Regulations.
3. Factories Act, 1961. Memorandum by the Senior Electrical Inspector of Factories—Deals with installations in factories.
4. Explanatory Notes on the Electricity Supply Regulations, 1937— These indicate the requirements governing the supply and use of electricity.
5. Hospital Technical Memoranda No. 7—Indicates the electrical services, supply and distribution in hospitals.

All electrical contractors are most particularly concerned with the various requirements laid down by Acts of Parliament (or by Orders and Regulations made thereunder) as to the method of installing electric lines and fittings in various premises, and as to their qualities and specifications.

Statutory Regulations:

1. Building (Scotland) Act, 1959—Provides for minimum standards of construction and materials including electrical installations.

2. Building Standards (Scotland) Regulations, 1967—Contains minimum requirements for electrical installations.
3. Electricity Supply Regulations, 1937—Indicates the requirements governing the supply and use of electricity.
4. The Electricity (Factories Act) Special Regulations, 1908 and 1944—Deals with installations in factories.
5. Coal and other Mines (Electricity) Regulations, 1956—Deals with installations at mines of coal and fireclay.
6. The Cinematograph (Safety) Regulations, 1955—Deals with installations in cinemas.
7. The Quarries (Electricity) Regulations, 1956—Deals with installations at quarries.

The British Standards Institution (BSI) is the approved body for the preparation and issue of standards of testing the quality of materials, and their performance and dimensions. The Codes of Practice which are also issued by the BSI indicate standards of good practice which take the form of recommendations. Some of the Codes which are of interest to the electrician include:

CP 327.402	Staff location systems
CP 327.403	Impulse clock and timing systems
CP 1003	Electrical apparatus for use in explosive atmospheres
BS 5405	Maintenance of electrical switchgear
CP 1011	Maintenance of electric motor control gear
CP 1013	Earthing
CP 1017	Distriubution of electricity on construction and building sites
CP 1018	Off-peak electric floor-warming systems
CP 1019	Installation and servicing of electric firm alarm systems
CP 1021	Cathodic protection

In order to ensure as far as possible that all electrical installation work carried out in this country is to a minimum standard, the National Inspection Council for Electrical Installation Contracting (NICEIC) was set up in 1956. The NICEIC is a non-profit-making body, supported by all sections of the electrical industry, for the protection of consumers against faulty, unsafe or otherwise defective electrical installations. Contractors on the Council's Approved Roll are those who have had their premises, equipment and installation work inspected and accepted as satisfactory by one of the Council's inspecting engineers. Thus, through its inspectorate, the Council is able to ensure a uniform standard of workmanship. The electricians employed, in whatever capacity, by an approved contractor are also, by association with their employer, accepted as being qualified and competent to carry out installation work.

4

Electricity supply and distribution systems

Though there are many forms of energy, it is generally accepted that electricity is by far the most convenient and flexible. The bulk of electricity generated in this country is obtained by burning a low-grade coal to produce high-pressure steam. This steam is then fed to machines known as steam-turbines, which are mechanically coupled to electric generators. For reasons of economy in fuel, plant and manpower electricity is generated in extremely large power stations which are sited throughout the country at places as close as possible to where the highest load demands occur.

The electric generators or alternators are three-phase machines generating alternating current. As the name implies, alternating current refers to a current whose direction of flow is first in one direction and then in an opposite direction, these alternations occurring at a regular period or frequency. One complete alternation is called a 'Hertz'. The frequency of mains electricity is standardised at 50 Hertz (Hz).

In certain parts of Britain, where there are large amounts of water-power, the water is harnessed to drive water-turbines. The site of a hydroelectric power station is very much determined by natural conditions, which are usually found to be most convenient quite long distances away from the heavy load centres. For the efficient operation of hydroelectric plant, there must be water available at a usable head (height above sea-level) and in sufficient quantity. And if the flow of water is not regular enough for a continuous supply for the turbines, then accommodation for water storage must be provided by building dams. The water from the dams is led through large-diameter pipes to the power stations.

Scotland and some parts of the mountainous areas of Wales have sufficient water-power resources for hydroelectric schemes. The bulk of hydroelectric generation is, however, in central and north Scotland. The turbines, as in steam power stations, are mechanically coupled to the electric generators.

The most modern method of generation is by means of nuclear power, in which the heat produced by nuclear fission is used to raise steam, which is then fed to steam turbines to drive the electric generators.

Other methods of generation in Britain include the diesel engine, very common for private standby plants, and in areas such as the small Scottish islands where there is no water-power, and where it would not be economic to take coal to them. In the Orkney islands, wind-power has been harnessed to experimental generators. And in the north of Scotland an unsuccessful attempt was made some years ago to make use of peat as a primary fuel for a steam station.

Whatever the method used to produce mechanical energy, the generator produces electrical energy generally at a voltage of 11,000–33,000 volts. This energy is then fed through the generator switchboards and circuit-breakers to transformers. As their name implies, the transformers change the voltage of the generators to higher values which are standardised at 132,000 volts (132 kV), 275kV and 400kV.

The reason for the transformation from a lower voltage to one considerably higher is that it is much more economical to transmit bulk supplies of electrical energy by using the highest voltage possible. In this way, overhead lines and underground cables need have only comparatively small conductors, with the minimum of electrical (I^2R watts) losses. To illustrate this aspect of electricity transmission, a conductor of 18 mm diameter is sufficient to transmit 50,000kW at 132,000 volts. To transmit the same amount of power at 250V, the conductor diameter would have to be something like 400 mm.

All electricity generated in this country under Statutory requirements is fed into what is known as the 'National Grid'. This is a vast and complex network of overhead lines and underground cables carrying power at high voltages to centres of high load-density. The history of electricity supply is interesting. The first power station was built in Deptford, London, by Ferranti in 1880. He tried to provide a high-voltage ac; he was not successful and this made supply engineers turn their attention to dc (direct current). Distribution was first at 110V; most of the generation of electricity was in the hands of private companies.

The electricity supply to consumers was rather primitive, though quite workable. One very successful system was devised by Crompton. He stretched bare conductors between glass insulators which were contained in undergound ducts or culverts. One such system lasted for some thirty years until, in 1926, it began to give trouble. This method was sufficient for the supply of energy at low voltages, but not satisfactory for high voltages. When it was realised that it was more economical to distribute energy at high voltages, the conductors were given some insulation and were known as cables. One of the earliest attempts to make a high-voltage, paper-insulated cable was by Ferranti. He wrapped oil-impregnated paper round copper rods and pushed the insulated conductors into lengths of iron pipes which he filled with compound.

As the demand for electricity grew, more independent power stations were set up throughout the whole country, either by private companies or by municipal authorities. Eventually the stage was reached where it was recognised that electricity was a real national asset and received the attention of the Government. In 1921 there was the proposal for a national grid system of high-voltage transmission lines which would interconnect all large generating stations. The Electricity Supply Act of 1926 established the Central Electricity Board. This Board was made responsible for the generation of electrical energy in England and Wales. The arrangement in Scotland was separate. In 1948, the whole business of electricity supply was taken over by the British Electricity Authority. The Board was to be responsible for the statutory generation and distribution of electricity in Great Britain. Further developments have led to the present-day set-up which consists of the Electricity Council and the Central Electricity Generating Board (CEGB) which took over the functions formerly exercised by the Central Electricity Authority. The Council is the coordinating body for the whole supply industry in England and Wales and advises the Minister of Power on all matters affecting the industry. The CEGB owns and operates the power stations and the National Grid, and is responsible for the generation and transmission of bulk electricity to the Area Electricity Boards. The Boards are autonomous bodies responsible for the distribution and retail sale of electricity to consumers. In Scotland the general supervision of the industry is the responsibility of the Secretary of State for Scotland. The country is divided into two Regions: the North of Scotland Region under the North of Scotland Hydro-electric Board, and the South of Scotland Region under the South of Scotland Electricity Board.

The energy at high voltages transmitted by the National Grid system is fed into grid substations for the transformation of the transmission voltage to secondary transmission voltages of 132kV and 33kV. This secondary transmission is for large consumers such as factories and those in areas of high load densities. Further substations reduce the secondary transmission voltages to 11kV and 3·3kV. The term 'distribution' is usually used to refer to the feeding of electrical energy through overhead lines and underground cables to supply small industrial, commercial and domestic premises.

For many years, the mains supply was dc. But gradually as the advantages of ac became apparent, there was a changeover to the latter supply. The voltage of the supply provided by the supply authorities varies according to the consumer and his needs.

Figure 4.1 summarises the pattern of electricity generation, transmission and distribution in Britain today.

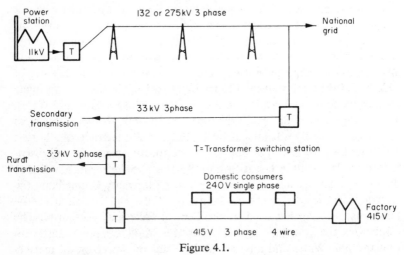

Figure 4.1.

If a supply authority agrees to give a supply to a consumer it must comply with the requirements of the Electricity Supply Regulations. One Clause in particular (No. 34b) is of particular interest. This states that the supply authority must constantly maintain the type of current (dc or ac), the frequency and the declared voltage. The frequency, however, may be varied by one per cent; the declared voltage can be allowed to vary by 6 per cent. These percentages are necessary to the supply authorities because variations in loads and voltage drops in cables would otherwise make the supply of electricity at inflexible voltages and frequencies quite uneconomic.

The Electricity Supply Acts define the voltages of supplies:

Extra-low voltage—30V (ac) or 50V dc

Low voltage—250V or less

Medium voltage—between 251V and 650V

High voltage—between 651V and 3,000V

Extra-high voltage—exceeding 3,000V (3kV)

Though today dc supply systems are not common, a word or two about them will be of interest. The early dc system was two-wire, with a positive and negative conductor. The supply voltage varied between 110V and 250V. As the need to transmit larger quantities of energy increased, a new system of distribution was adopted: the three-wire system. This consisted of a generator generating, say, 500V. A positive conductor and negative conductor ('the outers') provided a voltage of 500V between them. A third conductor called the neutral or 'mid-wire' was earthed and thus provided a supply voltage of 250V between any outer and the neutral. Thus, a 500V motor could be supplied from across both outer conductors; and a 250V lighting load could be connected across either positive and neutral or negative and neutral. Machines called balancers were used to maintain the voltages in the three-wire system constant at all conditions of load. While working on a three-wire dc system of supply, it was always necessary to remember that the 'live' conductor could be either positive or negative in polarity—very important when connecting up battery-charging equipment.

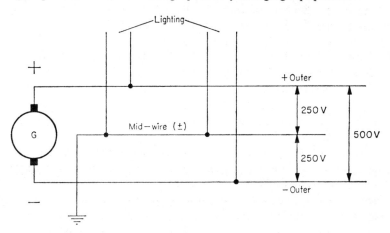

Figure 4.2. The simplified three-wire dc system.

There are three standard systems of ac supply. The first is an arrangement which gives two voltages, similar to the dc three-wire system. This ac system is known as three-phase four-wire and is illustrated in Figure 4.3. The supply is standard at 50Hz. There are three 'live' conductors called 'phases' or 'lines'. The voltage between any of these three phases is usually 415V. If a neutral conductor is earthed, then it will be found that the voltage between any phase conductor and the

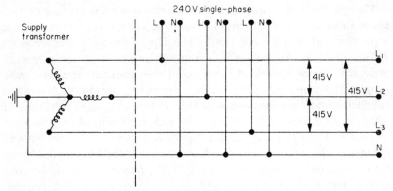

Figure 4.3.

neutral will be 240V. The three phases are identified by colours: red, yellow and blue; the neutral is always black. Supplies to premises are always connected to different phases to balance the loads. If the consumer is a small one, a house for instance, the supply cable contains two conductors, a live and neutral; the colour of the live will depend on the phase from which it has been taken. The supply voltage is 240V, and the system is known as single-phase, two-wire.

Larger consumers receive three-phase, four-wire supplies. The higher voltage is used generally for motors; lighting loads are connected across the outers and the neutral in such a way that when the whole installation is operating, the load across the three phases is reasonably balanced; that is, so that any one phase conductor does not carry a greater current than the other two.

Regulation No. 28a of the 1937 Electricity Supply Act states that energy for use at low voltage must not be introduced by more than one pair of conductors of a 3-wire or multi-phase system, unless the connected loads exceeds 8kW, *and* it is necessary to do so in order to avoid excessive volt drop or voltage variation on the distribution system in the vicinity. This Regulation does not affect current required for motors or other apparatus which requires medium voltage.

Now a brief word about the conductors used to distribute electrical energy to consumers. First, overhead lines. These are used because transmission by this method is very much cheaper than by underground cables. Indeed, in sparsely-populated rural areas, the cost of distribution can be kept reasonably low only by the use of overhead lines. For the National Grid, steel pylons are used. Porcelain and glass insulators carry the conductors which are either copper or aluminium with steel cores for strength. Other materials such as cadmium-copper and phosphor bronze are sometimes used. The choice of the material depends on cost, the required electrical and mechanical properties and on local conditions.

Small lines for rural supply at low and medium voltages use wood or steel poles and have short spans compared with the quite long spans between the Grid pylons. Creosoted wood poles are most common in this country. Insulators are nearly always made from porcelain (with a glazed finish which is self-cleaning). Conductors are generally of stranded copper and uninsulated, but those conductors which run from a pole to a house are insulated.

The type of cable used for supply depends on the voltage of the system. Below 1000V, the insulating material is either impregnated paper, varnished cambric, vulcanised rubber or a plastic material (PVC). Conductors are either stranded copper or aluminium. The protective sheath is lead or PVC. And the cables are usually armoured; that is, supplied with a sheath of stranded steel wire or steel tape. Cables designed for high voltages are usually paper-insulated and have, as might be expected, thicker insulation.

The general pattern of low- and medium-voltage distribution in towns and cities is for high-voltage supplies to be taken into a substation, which contains transforming equipment and switchgear. Also included in the circuitry are the protective apparatuses which ensure that faults in cables are detected and prevent damage to equipment and the cables. From these substations are taken cables which supply feeder pillars— these are large boxes so situated that a number of streets can be fed through fuses. The cables are known as feeder cables, from which the service cables (to premises) are connected.

In rural areas, extensive use is made of pole-mounted transformers. The high-voltage input is taken from the main overhead lines. The output conductors from the transformer are sometimes insulated. Where the supply is three-phase, four-wire, there are four conductors. In areas where a good earth is not easily obtainable, a fifth wire, a continuous earth wire (CEW) is provided by the supply authority for its consumers.

The sizes of service conductors (the conductors feeding the average domestic or small-consumer installation) is generally 16mm². This is reckoned to be sufficient for a current of about 70A, quite enough to cater for the electrical requirements of a house or flat with a floor area not exceeding 100 m².

When all the initial details about supply voltage and frequency and number of phases have been settled between the supply authority and the consumer, the form of tariffs must be agreed on. For the normal consumer there are two types of tariff: the ordinary tariff and the off-peak tariff. However, there are many others, depending on the type of consumer.

The domestic tariff applies to premises used exclusively as private dwelling houses. The ordinary tariff is usually based on a block of units (primary units) at a relatively high rate, after which the price per unit (the secondary units) drops to a low figure of about an average of 2p. An example of such a tariff is:

For each first 90 primary units consumed each quarter—4.18p.
For each additional unit consumed each quarter—1.62p.

Sometimes there is a minimum charge of about 65p per quarter for the installation of a prepayment meter.

Where the premises are other than private houses, a boarding house for instance, the general block tariff may apply, where a block of, say, the first 150 units used per quarter are charged at 4.18p; for each additional unit consumed each quarter the charge is 2p.

The off-peak tariff generally is available for approved supplies to certain types of apparatus which are disconnected from the mains at times specified by the local electricity board. The apparatus must be so arranged that the circuit is separate from others in the installation. For the purpose of disconnecting the apparatus (usually heating) special equipment is necessary, usually controlled by a time-switch supplied by the authority. The off-peak periods are generally from 1900 hrs to 0700 hrs, with two or three hours in the afternoon, and all weekend.

The industrial-power tariff, usually on a monthly or quarterly basis, applies to premises used for the purposes of manufacture. Again tariffs vary widely. But generally the consumer must pay for a maximum demand as follows:

There is a fixed charge per kilovolt-ampere (kVA) or kilowatt (kW) based on the maximum reading as measured by a Maximum Demand Indicator. Thereafter the charge per unit is for the actual energy consumed. Sometimes blocks of units apply. Charges vary also according to the voltage of the supply, whether medium or high voltage.

The commercial power tariff is much the same as above, and applies to those premises used for business rather than manufacture.

Farms, public lighting and commercial cooking are also given special tariffs. Small loads are metered by either credit or prepayment meters. In the first type, the energy consumed by the installation is recorded by the meter, which is read each quarter: the consumer receives his bill each quarter. The prepayment method involves the consumer in inserting coins into the meter to pay for his energy before he uses it.

Large consumers with inductive loads, such as motors, are penalised for having a low power factor. To improve the power factor, capacitors are installed. They involve an initial capital cost, but save money after a short period. See Chapter 27 'Capacitors and power factor'.

The reason for tariffs is, of course, that electricity generation involves the expenditure of large capital sums. It is estimated that in this country the demand for electricity doubles itself every ten years. Thus new generating stations must be built, transmission lines and cables installed, and transformers and switchgear bought to cater for the increased energy demand. There are also maintenance costs, overheads, administrative costs and insurance. These costs are usually constant and do not vary according to the output of a station. The other area of cost is related to the running of plant which does vary with the output of the supply system. For instance, an increase in the cost of coal will lead to an increase in the cost per generated unit, which is generally absorbed partially by the authority and partially by the consumer.

The place of batteries as a means of supplying electric power must be mentioned. They are generally used for standby supplies which come into operation when there is a failure of the mains. They are essential in theatres and hospitals and other premises where continuity of the supply is essential.

The standby battery is charged from the mains (rectified current if the supply is ac). The changeover may be manual or automatic; if the lighting or the supply to other services fails, the battery supply takes over. Generally a battery supply is designed to last up to two hours. Among the battery-standby systems in use are normal-voltage secondary lighting, and low-voltage lighting.

5

Supply-control and distribution on premises

The supply-intake position on any premises, large or small, is generally a matter of agreement between the electrical contractor and the supply-authority's engineer. Some thought is given to the matter because the authority want the position to be in the most convenient place in the building, so that the cost of the supply equipment (cable, cutouts, and the like) is kept to a minimum. The electrical installation engineer, acting on behalf of his client, is also interested in agreeing on a suitable position so that the electrical installation can be planned with the greatest economy and facility. Not the least important factor, especially where the meter-reader is concerned, is the decision to put the supply-intake position in a place where meters and controls are easily accessible. Too often meters are put in the most awkward positions because if they were put elsewhere they would be unsightly and not acceptable to the house-holder. However, a suitable cupboard or hinged box can be supplied, which can then be painted or decorated so as to be hardly noticeable.

For small installations in towns and cities, the cable is usually two-core, PVC-insulated and provided with sheathing, steel-wire armouring and served with a protective covering of tarred jute or tape. The junction which this cable makes with the street-main is contained in a tee-box, generally buried under the pavement just outside the premises. The two-core service-cable conductors are jointed to two of the main cable cores: one to the neutral and the other to one of the phase conductors (red, yellow or blue). The connectors are soldered, using the usual cable tee-joint, or by using the quicker method of crimping.

In rural areas, with overhead-line distribution, the house service cables are connected to the line conductors by means of mechanical connectors called line-taps. Conductors to the premises are always insulated, and are in most instances PVC-insulated. The service cables are taken to insulators mounted on D-irons, cleated to the walls of the house, and then run to the supply-intake position.

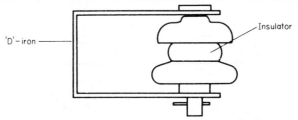

Figure 5.1. D-iron and insulator for overhead rural distribution.

Whether overhead or underground services, three-phase, four-wire connections are made in a similar manner to the two-wire services.

Whatever the size of the installation, there must be provision for effective control and protection. The IEE regulations indicate what is required to satisfy this requirement. The Regulations (Reg. A1) make it clear that every consumer's installation must be adequately controlled by specified equipment; and also that this equipment must be readily accessible to the consumer and that it is as near as possible to the supply-intake cutouts. The electricity regulation of the Factories Act says that 'efficient means, suitably located, shall be provided for cutting off all pressure from every part of a system, as may be necessary to prevent danger'.

The type and size of the main switchgear installed will, of course, depend on the type and the size of the installation and its total maximum load.

Thus, the main switchgear in any installation must be able to:

(*a*) isolate the complete installation from the supply;
(*b*) protect the installation against excess current, which may arise in, say, a short circuit;
(*c*) cut off the current should a serious earth fault occur, say a live conductor touching earthed metalwork.

The sequence of supply-control equipment in a single-phase installation is shown in the typical Figure 5.2a. The sequence for a three-phase, 4-wire installation is shown typically in Figure 5.2b.

The incoming cable can be underground or overhead. The cores are then prepared in the appropriate termination equipment. If overhead lines are used, there is a large main fuse cutout and generally a neutral connector-block. If the cable is underground, the termination is in the form of a sealing-box combined with a cutout unit. The sealing-box, always used with paper-insulated cables, serves to seal the end of the

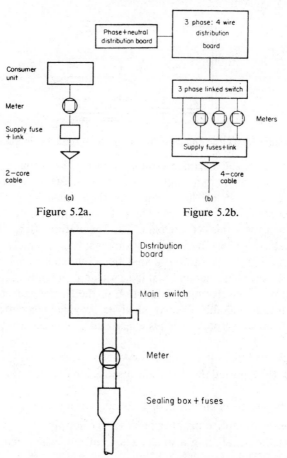

Figure 5.2a.   Figure 5.2b.

Figure 5.2c. Supply-intake position
with main switch and distribution
board.

cable to prevent any moisture from entering into the cable and possibly, in time, causing damage to the insulation. The sealing-box is filled with a pitch-tar compound. The cutout equipment contains the service fuse and neutral link. The rating of the fuse is such that it will carry indefinitely the maximum current that will flow when the total load is connected to the supply. For the normal domestic installation, the rating is 60A, though some installations which are 'all-electric' will require an 80A fuse. The fuse can be the rewirable or the cartridge type. The neutral link is a solid link generally of flat tinned copper, and is used where the neutral side of the supply is effectively earthed. If the neutral side of the supply is not earthed, then two fuses, one in each conductor, are provided.

Access to the cutouts is restricted to all except the supply authority's engineers. To prevent tampering by unauthorised persons, such as unscrupulous householders, the cutouts are sealed, and the seals must remain unbroken.

The next item in the sequence of supply-control equipment is the meter. In installations where a single tariff applies, one meter only is needed. Where, however, two or more tariffs apply, then the metering arrangements must cater for this. The purpose of any meter is to record the amount of electrical energy used by the current-using items connected to the electrical installation. The meter thus records the product of V (volts) and A (amperes) multiplied by t (time). The standard unit is the killowatt-hour (thus $250V \times 4A \times 1$ hour = 1 unit or 1kWh; or $250V \times 2A \times 2$ hrs = 1 unit or 1kWh). The meter terminals are sealed against tampering and unauthorised entry.

All the equipment so far described is the property of the supply authority.

From the meter the installation main cables are taken to the main switch or switchfuse, or consumer unit.

The consumer's main switch must be of the double-pole, linked-blade type which will isolate the complete installation from the supply when the switch is operated. If the supply is single-phase, both poles will be broken; if the supply is three-phase, or three-phase and neutral, then all three, or four, poles will be broken.

Fuse provision in the main switch is either a rewirable or cartridge fuse placed in the live conductor. In some instances the neutral may be fused, but more often it is a solid link. The fuse is the consumer's main fuse and is generally of the same rating as the service fuse, i.e. 60A. When the switch and fuse are combined in one unit, the item is known as a switchfuse. If only one tariff applies, then only one switchfuse will be required. If, however, there are two tariffs, then one switchfuse will be required for each tariff circuit. Switchgear must be fitted as close as possible to the main cutouts and meter, so that the cables from the meter (the 'tails') are as short as possible.

The distribution fuseboard (DFB) is the item which distributes the electricity to the various circuits which go to make up the complete electrical installation. In most DFBs, the circuits are provided with individual fuses, placed in the 'live' conductor; the neutral takes the form of a connector bar or block, thus maintaining the 'solid' link right through to the service cutouts.

DFBs are also available with circuit-fuses for both sides of the supply.

The 'splitter' unit is a combined mainswitch and distribution fuseboard. They are suitable for small lighting and heating installations where expense is a factor. However, the main disadvantage is that it is necessary to switch off all circuits before the lid can be opened to replace a fuse.

Instead of a DFB, an MCB-distribution board can be used. This item contains, instead of fuses, miniature circuit-breakers which give over-current protection to each individual circuit. The domestic consumer's control unit, or consumer unit, consists of a 60A main switch (which isolates both conductors) and a number of circuit fuseways. Use of this type of unit means that the consumer's main fuse may be omitted, provided permission is obtained from the supply authority.

The terms given to the circuits are important. Main cables are those which carry the total current of the installation, from the cutouts and meters to the main switch, through to the DFB. From the DFB are then taken subcircuits or final subcircuits. The definition of a final subcircuit indicates that such a circuit is not split up to supply or feed other circuits.

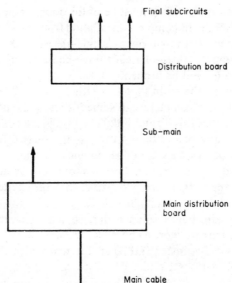

Figure 5.3. Sub-main and final subcircuit distribution.

Industrial installations, with a three-phase, 4-wire supply, often have rather more complicated arrangements at their supply-intake positions.

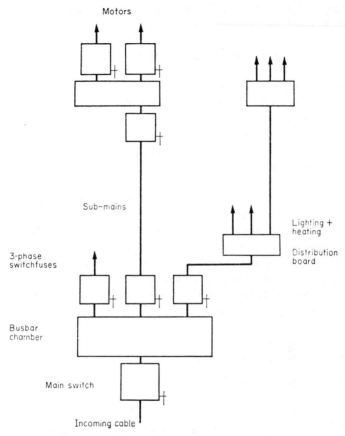

Figure 5.4. Typical distribution layout for a small industrial installation.

In multi-storeyed buildings, the supply (three-phase, 4-wire) is taken into a basement room. Then copper conductors of large cross-section are installed to form what are called 'rising mains'. Each flat or floor is tapped off these conductors. The arrangement of the supply-intake position for each consumer is then much the same as found in ordinary domestic premises.

All switchgear and control gear for any installation must be of suitable capacity. This means that the gear must be capable of carrying, without damage to itself or its associated wiring, the current that will flow when the installation is being operated in normal circumstances. IEE Regulations A21 and A22 say that 'diversity factor' may be applied when working out the size of conductors for main and submain cables, provided the conditions of operation are known.

This diversity factor can also be applied to switchgear. The Table in Appendix 1 of the IEE Regulations indicates how the diversity factor is applied to various types of electrical installations. The factor does not apply to final subcircuits. In one case, however, it is applied automatically: this is the 13A socket-outlet circuit, for details about which see Table A3 in the IEE Regulations.

A final subcircuit can range from a pair of 1·0 mm² cables feeding a light to a very heavy 3-core cable feeding a large motor from a circuit-breaker or switch at a main switchboard. The one important rule which applies to final subcircuits is that mentioned in the Electricity Supply Regulation 27: 'All conductors and apparatus must be of sufficient size and power for the work they are called on to do, and so constructed, installed and protected as to prevent danger.' The IEE Regulations ensure that this Regulation is complied with.

IEE Regulation A23 states that 'Every final subcircuit shall be connected to a separate way in a distribution board'. IEE Regulation A10, which applies to all final subcircuits, indicates that the current rating of a fuse for a final subcircuit shall not exceed the current rating of the lowest-rated cable (or flexible cord) in the circuit protected by that fuse. Certain exemptions to this rule are indicated in A10 Regulation.

There are five important general groups of final subcircuits:

(*a*) Rating not exceeding 15A.
(*b*) Rating exceeding 15A.
(*c*) Rated for 13A fused plugs.
(*d*) Rated for feeding fluorescent and other discharge-lamp circuits.
(*e*) Rated for feeding a motor.

(*a*) Table A2 in the IEE Regulations (and Regulation A25) indicates the final subcircuits which have a rating not exceeding 15A. The main requirements for such circuits is that the number of points which may be supplied is limited by their aggregate demand as determined by Table A2; there must be no allowance for diversity; and that the current rating of the cable must not be exceeded.

Domestic lighting circuits should be at least two in number. The reason for this recommendation is that should one circuit fuse, the

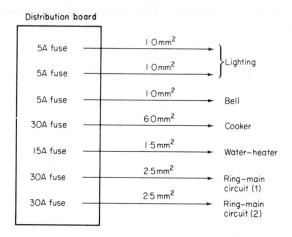

Figure 5.5. Typical final subcircuits from a distribution board.

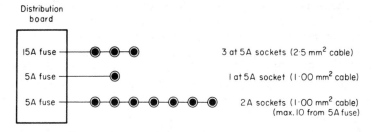

Figure 5.6. Distribution of socket-outlets of various ratings.

other circuit will supply the remainder of the lights in the house; that is, allow sufficient lighting to enable repairs to be carried out. In practice today domestic lighting circuits are rated at 5A. This means that they use 1·00 mm² cable, are protected by a 5A fuse, and the flexible cords of 0·75 mm², or 0·5 mm².

In industrial installations, it is usual to find that the wattage of lamps can be 300W, 500W and even 1000W. This means that a lighting circuit could well be rated at 15A. In this instance the cables would be 2·5 mm², 4·00 mm², and the rating of the circuit fuse 15A.

(*b*) Subcircuits with a rated capacity exceeding 15A include those feeding cooker units, and water-heaters which are rated at 15/20A. Final subcircuits rated at over 15A must not feed more than one point. There are two exceptions to this. One is where there is a socket-outlet integral with a cooker unit. The other applies to circuits feeding 13A socket-outlets, used with fused plugs.

(*c*) The first suggestion of a standard socket-outlet for domestic purposes and provided with a fused plug was made in 1944, in connection with a report on electrical installations in postwar buildings.

Up until the outbreak of the Second World War, the range of types of socket-outlets and plugs was so great that if one went from one part of the country to another, one had to replace all one's old plugs or appliances. As a solution to this problem, the standard socket-outlet and plug came into existence and is now accepted by virtually all installation engineers. Basically, the 13A circuits use socket-outlets with rectangular sockets. The appropriate plug has rectangular pins and contains a cartridge fuse the rating of which depends on the amount of current taken by the appliance connected to the plug.

IEE Regulation A34 deals with circuits for socket-outlets fitted with fused plugs and rated not more than 13A. Table A3 in the IEE Regulations indicates how these socket-outlets are to be connected. It will be noticed that a diversity factor has been applied. For instance, two socket-outlets can be wired with 2·5 mm^2 cables and protected by a 20A fuse. The reason for this seeming contravention of the Regulations is that it is unlikely that both 13A socket-outlets would ever supply their full rated current. This is even more the case in a ring-main circuit where some socket-outlets never supply more than is needed for a table lamp, a radio or television set (i.e. less than $\frac{1}{2}$A).

Figure 5.7 shows a typical ring-circuit arrangement which complies fully with the Regulations' requirements. Thirteen-amp socket-outlets with fused plugs should be used only on ac systems. The maximum number of 13A socket-outlets connected in any one circuit will depend on the estimated peak loading of the circuit. The total permissible load of fixed appliances fed from a single final subcircuit must not exceed 15A.

(*d*) One of the main requirements so far as the electrician is concerned with regard to fluorescent lighting circuits is the switching. In such circuits, it is not sufficient to take the total wattage of the lamps (as one would do with filament lamps, taking a minimum of 100W for any lampholder). Other factors must be considered. The inclusion of a choke or inductor increases the current taken by the circuit. IEE Regulation G3 requires that circuits shall be capable of carrying the total steady current, viz. that of the lamps and any associated gear, that is, lamp watts multiplied by not less than 1·8. This factor is based on the assump-

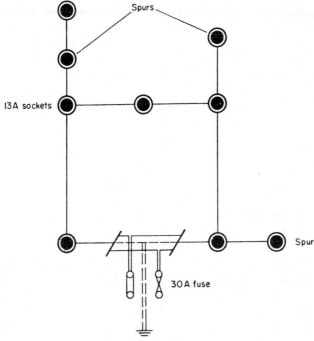

Figure 5.7.

tion that a circuit is corrected to a power factor of not less than 0·85, and takes into account control gear losses.

In the early days when fluorescent lighting was beginning to become popular, much trouble was experienced with the tumbler switches normally used for filament-lighting circuits. When they were switched off they often broke down between the live side and earth. This was due to the momentary high voltage induced by breaking an inductive circuit. IEE Regulation G2 says that any switch controlling circuits comprising discharge lamps should have a rating of not less than twice the steady current in the circuit it controls, unless the switch is specially designed to break an inductive load at its full rated capacity. The recommended switch-type is quick-make/slow-break.

(*e*) Final subcircuits which feed motors require some special consideration, though for the most part they must comply with the Regulations which apply to other types of final subcircuits. The current-rating of cables feeding a motor should be based on the full-load current taken by the motor. More than one motor may be connected to a 15A final subcircuit, provided that the total full-load rating of the motors does not exceed the rating of the smallest cable in the circuit. If, however, the rating of the motor does exceed 15A, then the circuit must supply one motor only. IEE Regulations A63 to A65 apply particularly to motor circuits, and mention the control of motors, overcurrent protection.

If any conductor carries a current, it will 'drop' a certain number of volts: $V = IR$, where I is the current and R is the resistance of the conductor. IEE Regulation B23 states that for domestic circuits, the maximum permissible volt drop must not exceed $2\frac{1}{2}$ per cent of the declared or nominal voltage. The drop is measured between the supply terminals (at the supply-intake position) and any, or every, point in the installation.

The *nominal* voltage is the voltage offered by the Supply authority and is not necessarily the voltage at the supply terminals, which may vary by ±6 per cent above or below the nominal voltage.

The volt drop on a 240V supply should not be greater than $2\frac{1}{2}$ per cent of 240 = 6V. Voltage drop in loaded cables is an important aspect of electrical installation work. For instance, if the voltage drops 10 per cent between, say, a DFB and a heater, then the heater output will drop by almost 20 per cent. Lighting is also affected, as are motors. It is thus not enough to choose a cable on the basis of its ability to carry the load current; the cable must also satisfy IEE Regulation B23.

The examples in Chapter 14, 'Lighting and power circuits', show the importance of choosing a cable to satisfy current and volt-drop requirements. Full familiarity of the Cable-rating Tables in the IEE Regulations will make these calculations easy to follow.

6

Conductors and cables

A 'conductor' in electrical work means a material which will allow the free passage of an electric current along it and which presents very little resistance to the current. If the conducting material has an extremely low resistance (for instance a copper stranded cable) there will, in normal circumstances, be no effect when the conductor carries a current. If the conductor material has a significant resistance (for instance, iron wire) then the conductor will show the effects of the electric current passing through it, usually in the form of an appreciable rise in temperature to produce a heating effect.

A 'cable' is defined as a length of insulated conductor (solid or stranded), or of two or more such conductors, each provided with its own insulation, which are laid up together. The conductor, so far as a cable is concerned, is the conducting portion, consisting of a single wire or of a group of wires in contact with each other.

The practical electrician will meet two common conductor materials extensively in his work: copper and aluminium.

As a conductor of electricity, copper has been used since the early days of the electrical industry because it has so many good properties. It can cope with onerous conditions. It has a high resistance to atmospheric corrosion. It can be jointed without any special provision to prevent electrolytic action. It is tough, slow to tarnish, and is easily worked. For purposes of electrical conductivity, copper is made with a very high degree of purity (at least 99·9 per cent). In this condition it is only slightly inferior to silver. At the present time copper is at an economic disadvantage because of its price, which has ranged from about £250 per ton to over £800 per ton. This has meant that more stable-priced aluminium has been allowed to take over many of the applications previously given to copper.

Aluminium is now being used in cables at in increasing rate. Although reduced cost is the main incentive to use aluminium in most applications, certain other advantages are claimed for this metal. For instance, because aluminium is pliable, it has been used in solid-core cables. Aluminium was used as a conductor material for overhead lines about seventy years ago, and in an insulated form for buried cables at the turn of the century. The popularity of aluminium increased rapidly just after the Second World War, and has now a definite place in electrical work of all kinds.

The following table shows the properties of copper compared with aluminium:

	Aluminium	*Copper*
Density at 20° (g/cm³)	2·70	8·89
Melting point (°C)	659	1083
Electrical conductivity (at 20°C)	62	100
Specific Resistance at 20°C (in microhm-cm)	2·8264	1·7241

In the above table, copper is taken as annealed high conductivity and aluminium as having a 99·5 per cent purity.

Aluminium has an excellent resistance to corrosion, though simple precautions must be taken to avoid corrosion in particular situations.

Conductors

Conductors as found in electrical work are most commonly in the form of wire or bars and rods. There are other variations, of course, such as machined sections for particular electrical devices (e.g. contactor contacts). Generally, wire has a flexible property and is used in cables. Bars and rods, being more rigid, are used as busbars and earth electrodes. In special form, aluminium is used for solid-core cables.

Wire for electrical cables is made from wire-bars. Each bar is heated and passed through a series of grooved rollers until it finally emerges in the form of a round rod. The rod is then passed through a series of lubricated dies until the final diameter of wire is obtained. Wires of the sizes generally used for cables are hard in temper when drawn and so are annealed at various stages during the transition from wire-bar to small-diameter wire. Annealing involves placing coils of the wire in furnaces for a period until the metal becomes soft or ductile again.

Copper wires are often tinned. This process was first used in order to prevent the deterioration of the rubber insulation used on the early cables. Tin is normally applied by passing the copper wire through a bath containing molten tin. With the increasing use of plastics materials for cable insulation, there was a tendency to use untinned wires. But now most manufacturers tin the wires as an aid in soldering operations. Untinned copper wires are, however, quite common. Aluminium wires need no further process after the final drawing and annealing.

All copper cables and some aluminium cables have conductors which are made up from a number of wires. These conductors are of two basic types: stranded and bunched. The latter type is used mainly for the smaller sizes of flexible cable and cord. The solid-core conductor (in the small sizes) is merely one single wire.

Most stranded conductors are built up on a single central conductor. Surrounding this conductor are layers of wires in a numerical progression of 6 in the first layer, 12 in the second layer, 18 in the third layer and so on. The number of wires contained in most common conductors are to be found in the progression 7, 19, 37, 61, 127.

Stranded conductors containing more than one layer of wires are made in such a way that the direction of lay of the wires in each layer is of the reverse hand to those of adjacent layers. The flexibility of these layered conductors is good in the smaller sizes (e.g. 7/0·85 mm) but poor in the larger sizes (e.g. 61/2·25 mm).

When the maximum amount of flexibility is required, the 'bunching' method is used. The essential difference of this method from 'stranding' is that all the wires forming the conductor are given the same direction of lay. A further improvement in flexibility is obtained by the use of small-diameter wires, instead of the heavier gauges as used in stranded cables.

When more than one core is to be enclosed within a single sheath, oval and sector-shaped conductors are often used. These shaped conductors are shown in Figure 6.1.

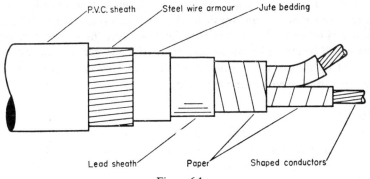

Figure 6.1.

It is of interest to note that when working out the dc resistance of stranded conductors, allowance must be made for the fact that apart from the central wire, the individual strands in a stranded conductor follow a helical path—and so are slightly longer than the cable itself. The average figure is 2 per cent. This means that if a stranded conductor is 100 m long, only the centre strand is this length. The other wires surrounding it will be anything up to 106 m in length.

Because aluminium is very malleable, many of the heavier cables using this material as the conductor have solid cores, rather than stranded. A saving in cost is claimed for the solid-core aluminium conductor cable.

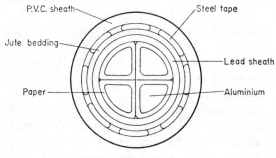

Figure 6.2.

Conductors for overhead lines are often strengthened by a central steel core which takes the weight of the copper conductors between the poles or pylons. Copper and aluminium are used for overhead lines.

Conductor sizes are indicated by reference either to their cross-sectional area (csa) or the number of strands and the diameter of each strand. Thus 7/0·85 mm is a conductor containing seven strands of wire, each wire having a diameter of 0·85 mm. The nominal csa of 7/0·85 to mm is 4·00 mm^2. It is important for both systems of cable nomenclature to be remembered because many of the larger sizes of cable are named by the csa figure only. And, of course, solid-conductor cables can only be named by the csa system.

Insulators

Many materials are used for the insulation of cable conductors. The basic function of any cable insulation is to confine the electric current to a definite path; that is, to the conductor only. Thus, insulating materials chosen for this duty must be efficient and able to withstand the stress of the working voltage of the supply system to which the cable is connected. The following are some of the more common materials used for cable insulation:

Rubber. This is one of the most common forms of insulation, and is used in the 'vulcanised' form, which is pure rubber with up to 5 per cent sulphur: it is flexible and impervious to water. The amount of mechanical strength it has depends on the degree of vulcanisation. And it will retain its good insulating properties where there is neither strong light nor undue heat. The maximum temperature in which vulcanised-rubber-insulated (VRI) cables should be installed is 55°C. Because the sulphur in the rubber attacks pure copper, VRI is always found used with tinned-copper conductors.

Butyl rubber. This material is a wartime development which is now being used increasingly for insulating cables designed to work in temperatures up to 80°C. The material has a slightly less mechanical strength than has natural rubber. And it tends to lose its resilience rather more quickly at very low temperatures. These are not serious disadvantages so far as electrical cables are concerned. Butyl rubber is less affected by water than any other natural or synthetic rubber compound. It retains its physical and electrical qualities after long periods of heating at temperatures which would completely destroy natural rubber. It also has excellent resistance to oils and chemicals.

Silicone rubber. This material can be used in operating temperatures up to 140°C, and is useful for many of the enclosed lamp fittings being installed at the present time. The physical properties of this material are such that some suitable protection is needed—usually glass braiding with a heat-resisting lacquer, though this is rather expensive. Terylene braiding is used as a cheaper, but just as ideal, substitute.

Polyvinylchloride. This plastic material is more commonly called 'PVC'. It was given a place in cable insulation during the Second World War when rubber was in short supply. The popularity of PVC has increased since then until the present time when it seems to be overtaking rubber. The properties of PVC are generally similar to those of rubber, though it tends to soften when it is installed in working temperatures higher than the recommended 65°C. At very low temperatures it tends to crack. Because it is virtually unaffected by chemical action, it is in some ways a better insulation material than rubber. It can withstand oils, and many types of acids and alkalis. However, it does not recover easily from stretching (as rubber can) and its insulation resistance is very much lower than that for rubber.

Paper. Paper has been used as a cable-insulation material since the very early days. The paper, however, must be impregnated with mineral oil or some other suitable compound. This must be done because paper is 'hygroscopic', that is, it absorbs moisture when in a dry state. When using impregnated paper for cable insulation, care must be taken to ensure that the impregnation is not dried out when operating in high temperatures (75°C).

Mineral insulation. This is composed of magnesium oxide powder and is used in the type of cable known as MIMS (Mineral-insulated, metal-sheathed). It has a very high degree of resistance to fire. The voltage rating of cables using this insulation is limited to 660V. The cable can withstand severe mechanical treatment during erection and operation and, because the insulation is not affected by high temperatures, it can be operated at temperatures up to dull red heat without affecting its properties. Temperature limits depend on the termination used—see Table B3 in the IEE Regulations.

Polychloroprene. This is another plastic material (PCP) which is used for the insulation of cables intended for operation in conditions which would harm both VRI and PVC. The material is resistant to oil and petrol, and can be used where there is exposure to sulphur fumes, ammonia fumes, steam, lactic acid, heat (limit is 55°C) and direct sunlight. PCP-sheathed cables are very suitable for farm installations.

Glass insulation. This material is very heat-resistant and is used for temperatures as high as 180°C. As glass-fibre, the insulation takes the form of impregnated glass-fibre lappings, with impregnated glass-fibre braiding. This insulation is found commonly in the internal wiring of electric cookers or other appliances where the cable must be impervious to moisture, resistant to heat and be tough and flexible.

Asbestos. This material is heat-resistant, but in an untreated state is not a good insulant. Suitably processed, however, it provides a cable covering with outstanding mechanical, thermal and electrical characteristics. The process involves thoroughly saturating the asbestos fibres with a chemically-neutral compound and compressed to form a uniform felted wall. In cables used for furnaces, boiler-rooms and so on, protection is afforded by lead-alloy sheaths.

Protection

Sheathing. Only in exceptional circumstances does the insulation of a conductor offer some protection against attacks by water, oils, acids and mechanical damage. Thus, it is common practice to protect the insulated conductor by a sheath or covering of some material which will enable the cable to be used in situations where some physical damage might result.

The basic purpose of the sheath is to prevent moisture from reaching the insulated core of the cable when in service. This implies that the sheath be impervious and resistant to corrosion. Once applied, a sheath must be sufficiently pliable to withstand a number of coiling and straightening operations during cable installation. Sheathing materials vary considerably, and are usually associated with the type of material used for the conductor installation. Rubber-insulated conductors have either tough-rubber (TRS) sheathing or lead-alloy sheathing (LAS). Where the cable is to be installed out-of-doors, the TRS is strengthened by a semi-embedded braiding and coated with compound. PVC-insulated conductors are sheathed with the same material. Mineral-insulated conductors are enclosed within a metal sheath which can be copper (MICS) or aluminium (MIAS). Paper-insulated cables generally have a lead-alloy sheath. Aluminium conductors are used with aluminium sheaths.

In many instances, the metal sheathing and armouring of cables are used to act as a conductor for earth-leakage currents.

Sometimes the wiring system acts as a sheath to protect against damage to the cables. For instance, conduit protects VRI- or PVC-insulated cables and the cables need not be provided with a sheath.

Armouring. In certain circumstances it is necessary for a cable to be protected against the occurrence of mechanical damage. Protection by 'armouring' is defined as the provision of a 'helical' wrapping or wrappings of metal (usually wires or tapes), primarily for the purpose of mechanical protection. The type of damage against which the cable is protected is rough treatment, abrasion, collision. The materials used, in tape or wire form, for armouring cables is most often steel. But aluminium is also used.

Special cables and conductors

There are many applications for cables and conductors, apart from the most common use, which is wiring to form part of an installation for lighting and current-using apparatus. Some of the most common types will be mentioned here, but the practical electrician may come across many other applications and types.

Possibly the most familiar conductor is that used for extra-low voltage work, for bells and other similar signalling applications. The feature of this conductor is the relatively thin covering of insulation—because the circuit would never be supplied with anything greater than extra-low voltage (see IEE Regulations Definition: 50V ac or 100V dc). The current carried is small.

Another type of conductor is called 'fittings wire', which is suitable only for the internal wiring of fittings with small bore aperture and not subject to disturbance or mechanical damage. The current carried is quite small.

Also of small csa are the telephone cables used not only for GPO work but also for the internal communication systems in offices and works; these systems are often installed by a contracting electrician with experience in this particular field.

Winding wires are yet another type of conductor used for a particular application. In this instance they are used for making the coils of electro-magnets and solenoids. The conductor material is nearly always high-conductivity copper with various insulating materials applied. The type of insulation depends on the application of the coil, and is generally one of three groups: paper, textiles and enamels. For transformer windings paper is used. In the textile group fall cotton, silk (natural and synthetic) and glass-fibres. Various types of varnished cloth are also to be found. In the enamelled-wire field there are types of enamels ranging from ordinary applications to those involving working temperatures of above 180°C.

The following are a few of the applications for winding wires formed into coils of one sort or another: small transformers, chokes, relays, solenoids; motor and transformer windings; electronic devices.

Soil-warming conductors are generally supplied with extra-low voltage and are bare specially-selected galvanised steel wires. The gauge is from 2·6 mm to 1·6 mm, the larger gauges being more resistant to corrosion and mechanical damage. For small domestic soil-warming applications, the circuit is fed from a double-wound transformer supplying 4 to 8V. For larger installations the voltage is up to 30V. The wires are really resistive conductors which dissipate heat when a current passes through them.

The elements of electrical heating apparatus are yet another group of conductors, possessing sufficient resistance to produce an effect: heat. Most element conductors are made from nickel-chromium, or nickel-iron-chromium, though other alloys are used depending on the application, and usually the final temperature required.

One now-common application of the resistive conductor or element is for surface heating; when provided with some suitable insulation they become heating cables. The applications include maintaining an easy-flow temperature of liquids in pipes, for heating jackets and soil-warming. An example of the latter is a central copper conductor, PVC-insulated, spiralled with a resistance heating element. A braiding of tinned-copper wire is provided and the whole is sheathed with PVC. These cables are designed for use on normal mains voltage (200V–250V) to eliminate the need for transformers for reducing the mains voltage to 8V–30V. The cables are manufactured to give so many watts per metre length.

Cable types

The range of types of cables used in electrical work is very wide: from heavy lead-sheathed and armoured paper-insulated cables to the domestic flexible cable used to connect a hair-drier to the supply. Lead, tough-rubber, PVC and other types of sheathed cables used for domestic and industrial wiring are generally placed under the heading of power cables. There are, however, other insulated copper conductors (they are sometimes aluminium) which, though by definition are termed cables, are sometimes not regarded as such. Into this category fall those rubber and PVC insulated conductors drawn into some form of conduit or trunking for domestic and factory wiring, and similar conductors employed for the wiring of electrical equipment. In addition, there are the various types of insulated flexible conductors, including those used for portable appliances and pendant fittings.

The main group of cables is 'flexible cables', so termed to indicate that they consist of one or more cores, each containing a group of wires, the diameters of the wires and the construction of the cables being such that they afford flexibility.

Single-core. These are tinned copper wires VR-insulated and braided and compounded, or VR-insulated and sheathed with tough rubber, or VR-insulated and PCP-sheathed, or VR-insulated and lead-sheathed, or lastly, PVC-insulated and sheathed. Sizes range from 1/1·13 mm to 127/2·52 mm, depending on the type of insulation. Other single-core cables include HSOS which are VR-insulated, varnished-calico taped, and braided and compounded with weather-resisting material. There is also the TR-sheathed and braided farm-wiring cable. Single-core cables are circular in section.

Two-core. Two-core or 'twin' cables are flat or circular. The insulation and sheathing materials are those used for single-core cables. The circular cables require cotton filler threads to gain the circular shape. Flat cables have their two cores laid side by side.

Three-core. These cables are the same in all respects to single- and two-core cables except, of course, they carry three cores.

Composite cables. Composite cables are those which, in addition to carrying the current-carrying circuit conductors also contain an earth-continuity conductor.

To summarise, the following groups of cable types and applications are to be found in electrical work, and which the electrician, at one time or another during his career, may be asked to instal:

Wiring cables. Switchboard wiring; farm wiring; domestic and workshop flexible cables and cords. Mainly copper conductors.

Power cables. Heavy cables, generally lead-sheathed and armoured; control. cables for electrical equipment. Both copper and aluminium conductors.

Mining cables. In this field cables are used for trailing cables to supply equipment; shot-firing cables; roadway lighting; lift-shaft wiring; signalling, telephone and control cables. Adequate protection and fire-proofing are features of cables for this application field.

Ship-wiring cables. These cables are generally lead-sheathed and armoured, and mineral-insulated, metal-sheathed. Cables must comply with Lloyd's Rules and Regulations, and with Admiralty requirements.

Overhead cables. Bare, lightly-insulated and insulated conductors of copper, copper-cadmium and aluminium generally. Sometimes with steel core for added strength. For overhead distribution cables are HSOS, PVC and PBJ (paper, braided-jute insulated) and in certain instances comply with Post Office requirements.

Communications cables. This group includes television down-leads and radio-relay cables; radio-frequency cables; telephone cables.

Welding cables. These are flexible cables and heavy cords with either copper or aluminium conductors.

Electric-sign cables. PVC- and rubber-insulated cables for high-voltage discharge lamps (neons, etc.) able to withstand the high voltages.

Equipment wires. Special wires for use with instruments, often insulated with special materials such as silicone, rubber and irradiated polythene.

Appliance-wiring cables. This group includes high-temperature cables for electric radiators, cookers and so on. Insulation used includes nylon, asbestos and varnished cambric.

Heating cables. Cables for floor-warming, road-heating, soil-warming, ceiling-heating and similar applications.

Flexible cords. A flexible cord is defined as a flexible cable in which the csa of each conductor does not exceed 56/0·30 mm. The most common types of flexible cords are used in domestic and light industrial work. The diameter of each strand or wire varies from 0·2 mm to 0·30 mm. Flexible cords come in many sizes and types; for convenience they are grouped as follows:

Twin-twisted. These consist of two single insulated stranded conductors twisted together to form a two-core cable. Insulation used is vulcanised-rubber and PVC. Colour identification in red and black is often provided. The rubber is protected by a braiding of cotton, glazed-cotton, rayon-braiding and artificial silk. The PVC-insulated conductors are not provided with additional protection.

Three-core (twisted). Generally as twin-twisted cords but with a third conductor coloured green, for earthing, lighting fittings.

Twin-circular. This flexible cord consists of two conductors twisted together with cotton filler-threads, coloured brown and blue, and enclosed within a protective braiding of cotton or nylon. For industrial applications, the protection is tough rubber or PVC.

Three-core (circular). Generally as twin-core circular except that the third conductor is coloured green and yellow for earthing purposes.

Four-core (circular). Generally as twin-core circular. Colours are brown and blue.

Parallel-twin. These are two stranded conductors laid together in parallel and insulated to form a composite cable with rubber or PVC.

Twin-core (flat). This consists of two stranded conductors insulated with rubber, coloured red and black, laid side by side and braided with artificial silk.

High-temperature lighting, flexible cord. With the increasing use of filament lamps which produce very high temperatures, the temperature at the terminals of a lampholder can reach 71°C or more. In most instances the usual flexible insulants (rubber and PVC) are quite unsuitable and special flexible cords for lighting are now being made. Conductors are generally of nickel-plated copper wires, each conductor being provided with two lappings of glass-fibre, silicone varnished and braided with glass-fibre. The braiding is also varnished with silicone. Cords are made in the twisted form (two- and three-core) and in the circular form (two- and three-core).

Flexible cables

These cables are made with stranded conductors, the diameters being 0·3 mm, 0·4 mm, 0·5 mm, and 0·6 mm. They are generally used for trailing cables and similar applications where heavy currents (IEE

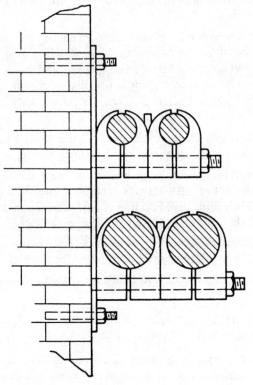

Figure 6.3.

Regulations, Table 20M ranges from 42A to 680A) are to be carried, for instance, to welding plant.

Cable installation
The installation of cables is dealt with fully in Chapter 12, 'Installation methods'. However, one or two relevant points can be made here, particularly with regard to cable supports, the identification of conductors, and cable bends.

Table B2 in the IEE Regulations indicates how the supports for electric cables are regulated with regard to the distance between them. In general, the heavier the cable, then the more support there should be per unit length (metre). The spacing is different for vertical and horizontal cable runs, the shorter spacing being for the horizontal runs. Cable supports include clips made from metal and in the form of buckles, or cleats, generally porcelain for smaller types of cables to claw-type cleats for the larger cables. Figures 6.3 and 6.4 shows a variety

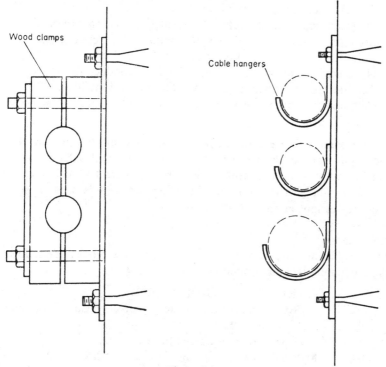

Figure 6.4.

of cable supports. Saddles are also used to hold single-core, two- and multi-core cables, and a group of single-core cables.

It is important to ensure that when a cable is bent the radius of the bend is not too small; that is the bend should not be sharp. Table B1 in the IEE Regulations indicates the minimum internal radius of bends for cables of different types, and considers the type of insulation, the covering (sheath and armouring) and the diameter.

Conductor identification

One of the most important aspects of wiring is the use of standard colours to identify cables and wires. Table B4A of the IEE Regulations indicates the colours used for bare conductors and the cores of cables, depending on the type of supply system to which the cables are connected. An alternative system of conductor identification is the use of numbers.

In general, conductor identification methods are as follows:

1. Colouring of the conductor insulation.
2. Printed numbers on the conductor insulation (e.g. on paper).
3. Coloured adhesive tapes at the terminations of the conductors.
4. Coloured sleeves at the terminations of the conductors.
5. Numbered sleeves at the terminations of the conductors.
6. Coloured paint for bare conductors (busbars and the like).
7. Coloured discs fixed near to the terminations of conductors (e.g. on a meter board).

As a point of interest, the British Standards Institution require that the colouring for cable identification should be fast, that is that it should not fade on prolonged exposure to light. For instance, PVC colours tend to fade in sunlight. The printing of numbers and letters must be such that they cannot be rubbed off in use.

Abbreviations

Many cables are known by their abbreviations and initials. The following are among the most common:

VRI:	Vulcanised-rubber insulated
CSP:	Chloro-sulphonated polyethylene compound (e.g. 'Hypalon')
TRS:	Tough-rubber sheathed
MIMS:	Mineral-insulated, metal-sheathed
MICS:	Mineral-insulated, copper-sheathed

NBR/PVC: Nitrile-butadiene Rubber/PVC blends
MIAS: Mineral-insulated, aluminium-sheathed
HSOS: House-service overhead system
PVC: Polyvinylchloride
PCP: Polychloroprene (e.g. 'Neoprene')
LAS: Lead-alloy sheathed
PILC: Paper-insulated, lead-sheathed
PILCSTA(SWA): Paper-insulated, lead-sheathed, steel-tape armoured (steel-wire armoured)
AVC: Asbestos varnished-cambric
SRI: Silicone-rubber insulated
BRI: Butyl-rubber insulated
PBJ: Paper-insulated, braided-jute
HCHD: High-conductivity, hard-drawn
VC: Varnished-cambric
PTP: Polyethylene terephthalate (e.g. 'Terylene'), usually varnished

7

Conductor joints and terminations

Joint. The connection of two lengths of conductor by a method which ensures a continuous path for the unimpeded flow of an electric current.

Termination. Applied to the end of a conductor prepared in such a way that it is suitable for connection to the terminal to which it is to be connected by mechanical means.

Basic electrical and mechanical requirements

The following are the basic requirements which must be met in any electrical connection.

1. There must be sufficient contact area between the two current-carrying surfaces (e.g. between wire and terminal). If this is provided, then the surface contact resistance will be minimised. There will also be a reduction in the voltage drop across the contacts and in the amount of heat generated. Note that the voltage drop is the product of the current (I) flowing through the joint or termination and the resistance (R) of the contact. The heat generated is calculated in watts and is the product of the square of the current (I^2) flowing through the joint or termination and the resistance (R) of the contact, thus I^2R watts. In practice, the volt drop and the amount of heat generated are so small that they are ignored. However, a badly-soldered joint (a 'dry' joint, for instance) could cause trouble and must be rectified before damage is done, particularly to any associated insulation.

2. There must be adequate mechanical strength. This aspect is very important where there is the possibility of leads being pulled. Thus, the type of conductor termination must be considered from the point of view of mechanical damage being sustained by the joint or termination.

3. The third requirement is the ease with which a connection can be made and unmade. Electrical wires are often 'permanently' connected by

soldering or crimping methods, usually where the currents to be carried are relatively low. Where, however, permanent connections are a disadvantage (e.g. in maintenance), then detachable unions are selected. These are invariably used in medium- and high-current work.

The resistance of two separable contact surfaces depends on the amount of pressure exerted to keep the surfaces together, and the conditions of the surfaces (e.g. uneven or dirty). Non-separable contacts soldered, brazed (or welded) depend on the effectiveness of the jointing method used to reduce resistance. The following are the main requirements of the IEE Regulations regarding terminations and conductor joints:

B60. *Cable terminations.* All terminations of cable conductors and bare conductors must be accessible for inspection. They must be electrically and mechanically sound. No stress should be imposed on the terminals. Where two dissimilar metals are being used (e.g. copper and aluminium), care must be taken to prevent corrosion, particularly in damp situations. All insulation damaged by heat-jointing processes (e.g. soldering) must be made good. Soldering fluxes which remain acidic or corrosive at the completion of a soldering operation must not be used.

B73. *Joints in cables.* Generally the requirements are the same as those indicated in Reg B60. An electrically sound joint means that the resistance of the jointed conductor should not be greater than that of an unjointed length of a similar conductor. A mechanically-sound joint means that any pulling on the finished joint will not disturb the joint. A soldered joint must be mechanically sound *before* soldering. A joint which is readily accessible is one which is located usually in a box of the inspection type and the box itself must, of course, be accessible. (B78). The termination of a flexible cable or a flexible cord to an appliance must be done either by wiring direct onto the appliance terminals or by means of an inlet connector (e.g. kettle connector). If a joint must be made between a flexible cord and/or flexible cable, an insulated mechanical connector must be used. Certain manufactured cable couplers and connectors are available on the market. They must be of a design in which no live part can be exposed in service. Flexible cables and cords must be so connected that no appreciable stress is placed on the conductors or the terminals.

F10/K26. Non-reversible cable couplers and connectors are desirable.

Joint Methods

The many methods used to join conductors may be reduced to two definite groups. The first group involves the use of heat to fuse together the surfaces of the joint (e.g. soldering and welding). The second group uses pressure and mechanical means to hold the surfaces together (e.g. clamping, bolting, riveting). The following are brief descriptions of the types of jointing method in each group.

Soldering

This jointing process is described in full in Chapter 38. Briefly, it involves the use of molten metal introduced to the two surfaces to be joined so that they are linked by a thin film of the metal which has penetrated into the surfaces. The metal used for joining copper surfaces is solder, which is an alloy of tin and lead. It melts at a comparatively low temperature. The grade of solder most suitable for electrical joints is tinman's solder (60 per cent tin, 40 per cent lead; melting point is about 200°C). The disadvantage of soldering is that it makes the joint a non-separable contact. Soldered joints in busbars must be reinforced by bolts or clamps.

Welding

This process is sometimes used for large-section conductors such as busbars. Welding is the joining of two metal surfaces by melting adjacent portions so that there is a definite fusion between them to an appreciable depth. The heat is supplied by a gas torch or an electric arc. Again, the welded joint is a non-separable contact.

Clamping

A clamped joint is easy to make, no particular preparation being required. The effective cross-sectional area of the conductor is not affected, though the extra mass of metal round the joint of termination makes a larger bulk. However, the joint or termination is cooler in operation. This method provides a separable contact. Surfaces must be clean and in definite mechanical contact. Precautions must be taken to ensure that the bolts and nuts of the clamp are locked tight.

Bolting

This method involves drilling or punching holes in the material and is more suitable for busbars. The holes tend to reduce the effective area of the material. Contact pressure also tends to be less uniformly distributed in a bolted joint than in one held together by clamps. Bolted joints can be dismantled easily.

Riveting

If well made, riveted joints make a good connection. There is the disadvantage, however, that they cannot easily be undone or tightened in service.

Crimping

This is a mechanical method. For conductor joints, a closely-fitting sleeve is placed over the conductors to be joined together and crimped (or squeezed) by a hydraulically- or pneumatically-operated crimping tool. Crimped lugs are also available for conductor terminations.

Mechanical connectors

These consist of one-way or multi-way brass terminals contained in blocks of porcelain, bakelite or plastic. Small screws are used to make the connection. The connector must be able to take all strands of the conductors to be joined together. Another type of connector is that used for joining flexible cords. Accessory connections are those used for terminating flexible cords or cables for connection direct to an appliance (e.g. iron, kettle).

Types of Joints

Scarfe Joint

This is one of the simplest joints to make and is used where the wires are not under any tensile stress and also where the joint is to be as small as possible. Though generally used with solid conductors, it can be used on stranded conductors provided the strands are soldered solid beforehand (See Figure 7.1).

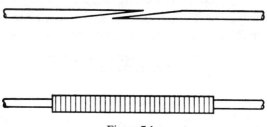

Figure 7.1.

Method

1. The end of each conductor is chamfered by careful filing for 40 mm.
2. Clean the ends. Tin them and fit them together so that the joint-length diameter is the same as the uncut conductor.
3. Hold the conductors tightly together and bind the joint with tinned binding wire over a distance of 65 mm.
4. Solder the joint. Wipe off any excess solder with a fluxed cloth-pad.

Britannia joint

This joint is used for single overhead copper conductors which are not under tension (see Figure 7.2).

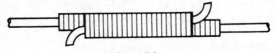

Figure 7.2.

Method

1. Thoroughly clean the ends of the conductors.
2. Tin the ends for a distance of about 75 mm.
3. Bend 6 mm of each end over at right-angles. This prevents the joint from pulling out when in use.
4. Bring the two wires together side by side over a distance of about 50 mm. The ends should face opposite directions.
5. Hold the joint in a small vice, or with pliers.
6. Bind both conductors together with $1\cdot00$ mm² tinned copper binding wire. The binding should be carried about 6 mm (about five turns) past each conductor end.
7. Solder the joint (the whole length of the binding). Wipe off any excess solder with a fluxed cloth-pad.

Bellhanger's joint

This joint is sometimes called a 'straight-twist' joint. It is not a strong type and so is used where no tensile stress is placed on the conductors, which are usually solid-core and insulated (see Figure 7.3).

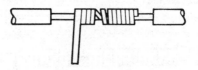

Figure 7.3.

Method

1. Remove the insulation from each conductor and for a distance of about 75 mm.
2. Clean the ends.
3. If VRI cable, strip off the braiding and tape for a further 12 mm on each conductor.
4. Lay the conductors together about 50 mm from the ends. Twist them tightly round each other in opposite directions. Each turn of a conductor should fit closely to the next turn. Pliers can be used to grip the crossed conductors. Each twist should contain about six turns.
5. Cut off surplus conductor ends on the bevel with side-cutters. Smooth over with pliers.
6. Solder the joint. Wipe of any excess solder with a fluxed cloth-pad.
7. Make good the insulation removed.

Telegraph joint

This joint is not strong and is generally used for single-core insulated conductors (see Figure 7.4).

Figure 7.4.

Method

1. Remove the insulation from each end for a distance of about 100 mm.
2. Clean and tin the ends.
3. If VRI cable, strip off the braiding and tape for a further 12 mm on each conductor.
4. Cross the conductors about 30 mm from the insulation, keeping the lefthand conductor in front.
5. Using pliers, grip the crossed conductors together.
6. Twist the lefthand end and the righthand main portion together. The turns need not be sharp.
7. Twist the righthand end and the lefthand main portion together.
8. Cut off surplus conductor ends on the bevel with side-cutters. Smooth over with pliers.
9. Solder the joint. Wipe off any excess solder with a fluxed cloth-pad.
10. Make good the insulation removed.

Tee-twist joint

This is a branch or 'tee' joint made with single-core conductors (see Figure 7.5).

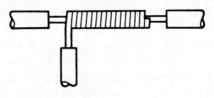

Figure 7.5.

Method

1. Remove the insulation from the through piece for a distance of 50 mm in the centre. This wire must not be cut.
2. If VRI cable, strip off the braiding and tape for a further 12 mm at each side.
3. Remove the insulation from the tee-piece for a distance of about 75 mm from the end.
4. If VRI cable, strip off the braiding and tape for a further 12 mm.
5. Clean the bared conductors and tin if necessary.
6. Place the tee-piece at right-angles to the lefthand side of the bared through-piece and tightly bind it round the through conductor from left to right.
7. Cut off surplus conductor end on the bevel with side-cutters and smooth over with pliers.
8. Solder the joint, leaving two or three turns free for flexibility. Wipe off any excess solder with a fluxed cloth-pad.
9. Make good the insulation removed.

Through-married joint

The married joint is used with stranded conductors. When made correctly it forms a very strong and satisfactory joint.

Method (3-stranded conductor) (See Figure 7.6).

1. Remove the insulation for 75 mm from each conductor end.
2. If VRI cable, strip off the braiding and tape for a further 12 mm on each conductor.
3. Twist the strands of each conductor firmly in the direction of lay for 25 mm, leaving 50 mm splayed out.

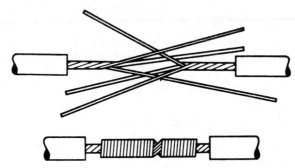

Figure 7.6.

4. Untwist the splayed strands and straighten them.
5. Interleave the strands and butt the twisted portions together. Each strand of one conductor should lie between two strands of the other conductor.
6. Hold down the strands of the righthand conductor along the lefthand conductor. Wrap the three strands of the lefthand conductor neatly around the righthand conductor half a turn at a time. Keep the strands tight and close together.
7. Wrap the strands of the righthand conductor round the lefthand conductor as in step 6.
8. Cut off surplus strands on the bevel with side-cutters. Smooth over with pliers.
9. Solder the joint. Wipe off any excess solder with a fluxed cloth-pad.
10. Make good the insulation removed.

Method (7-strand conductor)

1. Remove the insulation for 75 mm from the end of each conductor.
2. If VRI cable, strip off the braiding and tape for a further 12 mm on each conductor.
3. Twist the strands of each conductor firmly in the direction of lay for 25 mm, leaving 50 mm splayed out.
4. Untwist the splayed strands and straighten them.
5. Cut out the centre strand from each conductor.
6. Interleave the strands and butt the twisted portions together. Each strand of one conductor should lie between two strands of the other conductor.
7. Lightly bind the strands on the righthand side round the twisted strands.

8. Tightly bind the strands on the lefthand side round twisted strands against the lay of the conductor.
9. Untwist and straighten the strands on the righthand side. Bind them tightly round the twisted strands in the same direction as the twisted strands.
10. Cut off surplus strands on the bevel with side-cutters and smooth over with pliers.
11. Solder the joint. Wipe off any excess solder with a fluxed cloth-pad.
12. Make good the insulation removed.

Tee-married joint

Method (3-stranded conductor) (See Figure 7.7a).

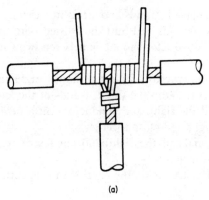

(a)

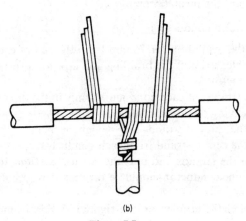

(b)

Figure 7.7.

1. Remove the insulation from the through-piece for a distance of about 30 mm in the centre. This wire must not be cut.
2. If VRI cable, strip off the braiding and tape for a further 12 mm at each side.
3. Remove the insulation from the tee-piece for a distance of about 75 mm from the end.
4. If VRI cable, strip off the braiding and tape for a further 12 mm.
5. Twist the strands of the through-piece.
6. Twist the strands of the tee-piece in the direction of lay for 25 mm from the insulation.
7. Untwist the remainder of length of the tee-piece and straighten strands.
8. Offer the tee-piece to the through-conductor so that two strands are made to go round the lefthand side of the through-piece and one strand round the righthand side of through-piece.
9. Tighten the strands in their respective directions.
10. Cut off surplus strands on the bevel with side-cutters. Smooth over with pliers.
11. Solder the joint. Wipe off any excess solder with a fluxed cloth-pad.
12. Make good the insulation removed.

Method (7-strand conductor) Figure 7.7b.

1. Remove the insulation from the through-piece for a distance of about 75 mm in the centre. This wire must not be cut.
2. If VRI cable, strip off the braiding and tape for a further 12 mm at each side.
3. Remove the insulation from the tee-piece for a distance of about 75 mm from the end.
4. If VRI cable, strip off the braiding and tape for a further 12 mm.
5. Twist the strands of the through-piece.
6. Twist the first 25 mm of the tee-piece in the direction of lay from the insulation.
7. Untwist remainder of length of tee-piece and straighten strands. Divide into 4 and 3.
8. Offer tee-piece to through-conductor so that four strands are over on one side and three strands on the other side of the through-conductor.
9. Tighten the strands of the tee-piece in respective directions round the through-conductor.
10. Cut off surplus strands on the bevel and smooth over with pliers.

11. Solder the joint. Wipe off any excess solder with a fluxed cloth-pad.
12. Make good the insulation removed.

Through-joint with conductor fitting

In order to save time and the cost of making married joints in cables, conductor fittings are used. For the through-joint, the weak-backed ferrule is used. It is a tubular piece of tinned copper opened along the top and weakened at the bottom by indenting. This allows it to be opened and closed easily.

Method (7-strand conductor) (See Figure 7.8a).

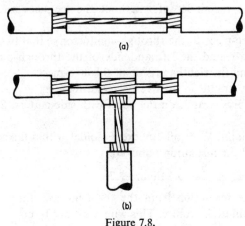

(a)

(b)

Figure 7.8.

1. Remove the insulation for 75 mm from the end of each conductor.
2. If VRI cable, strip off the braiding and tape for a further 12 mm on each conductor.
3. Place the ferrule on one cable. Put the cable ends together before tightening ferrule with pliers.
4. Solder the connector. Wipe off any excess solder with a fluxed cloth-pad.
5. Make good the insulation.

Tee joint with connector fitting

This type of joint uses a claw connector for tee joints in uncut through mains.

Method (7-strand conductor) (See Figure 7.8b).

1. Remove the insulation from the through piece for a distance of about 56 mm in the centre. This wire must not be cut.

2. If VRI cable, strip off the braiding and tape for a further 12 mm on each side.
3. Remove the insulation from the tee-piece for a distance of about 40 mm.
4. If VRI cable, strip off the braiding and tape for a further 12 mm.
5. Open out the claws of the connector and offer it to the through-conductor. Close the claws round the through-conductor using pliers.
6. Insert the end of the tee-piece into the connector. Tighten with pliers.
7. Solder the joint. Wipe off any excess solder with a fluxed cloth-pad.
8. Make good the insulation removed.

Mechanical connector through joint

This joint uses a porcelain-shrouded connector. (See Figure 7.9).

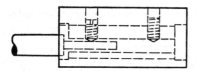

Figure 7.9.

Method (stranded conductors)

1. Remove the insulation from the end of each cable to leave 12 mm of exposed conductor.
2. If VRI cable, strip off the braiding and tape for a further 12 mm.
3. Twist the conductors together in the direction of the lay of the cable. Cut conductor to leave just over 6 mm exposed. If necessary, to fill the terminal hole, bend the conductor back on itself to leave just over 6 mm of conductor exposed.
4. Insert the conductor into the connector tube and tighten the grub screws securely using a suitable screwdriver.

Termination methods

There are many methods of terminating conductors for connection to accessories and current-using apparatus. The following is a short survey of some of the more common types of termination.

Punched and notched tabs

These generally accept a solid-core small-diameter conductor. The connection is soldered.

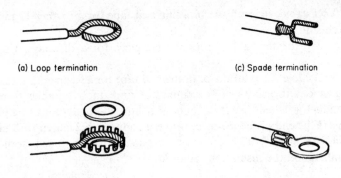

(a) Loop termination

(c) Spade termination

(b) Claw type termination

(d) Crimped termination

Figure 7.10. Types of conductor termination

Screwhead connection

The end of the conductor is formed into an eye using round-nosed pliers. The eye should be slightly larger than the shank of the screw, but smaller than the outside diameter of the screwhead, nut or washers. The eye should be so placed that the rotation of the screwhead or nut tends to close the joint in the eye. If the eye is put the opposite way round, the rotation of the screwhead or nut will tend to untwist the eye to make a bad, inefficient contact. Sometimes saddle washers are used to retain the shape of the eye.

Claw-type terminals

These are sometimes called segmented eyelet lugs. The conductor strands are twisted together tightly and formed into a loop to fit snugly into the circular claw. An associated brass or tinned copper washer is then placed on top. The claws are then bent over the washer.

Spade terminals

These are either preformed terminals, or made from the conductor end as follows (7-strand conductor):

1. Strip off a suitable amount of insulation from the end of the conductor. If VRI cable, strip off the braiding and tape for a further 12 mm.
2. Take one outer strand and twist round base immediately above the insulation.
3. Separate conductor into two sets of three strands each.
4. Twist each set tightly together.
5. Form a spade end.

Lug terminals

These come in many types as shown in Figure 7.11.

Connection between conductor end and the terminal's socket is made either by soldering or crimping.

Crimping. Select the correct terminal end. Strip the insulation from the cable end. Insert the wire into the open socket end of terminal and crimp using a crimping tool.

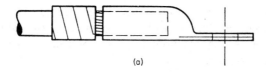

(a)

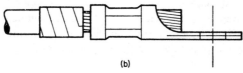

(b)

Figure 7.11.

Soldering. (1) Strip the insulation back about 50 mm. (2) Tin the socket of the lug. (3) Smear both the inside of the socket and the conductor end with flux. (4) Fit the socket to the conductor. If the socket is too large, the conductor diameter should be enlarged with a tinned-copper wire binding. (5) Play the flame of a blow-torch on the socket until the heat has penetrated to the conductor. Apply solder to the lip of the socket. (6) When the termination has cooled, cut back any damaged insulation and make good. Tape can be used to protect the original insulation.

A file should never be used to smooth or clean up a soldered connection. The solder should be smoothed by wiping it with a fluxed cloth-pad while the socket is still warm.

Line-taps

These are used for making non-tensioned service or tee connections to overhead lines. They are available in a range of sizes suitable for copper conductors. A simple shroud is provided to insulate the line-tap when used on covered service cable. There are designs for use with

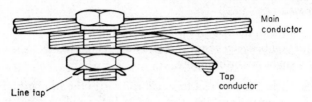

Figure 7.12. Typical line tap.

aluminium conductors and for bimetallic connections between aluminium and copper conductors. In these instances, the shroud is filled with
weatherproof sealing compound, giving protection against climatic
attacks and corrosion.

Joints and terminations on MICS cable

This type of cable consists of conductors insulated with compressed
magnesium oxide and enclosed in a seamless copper sheath or tube.
Generally, the ends of the cable must be sealed against the ingress of
moisture by using a suitable insulated sealing compound. The complete
cable termination (see Figure 7.13) comprises two subassemblies, each

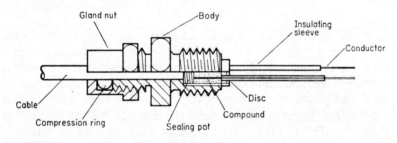

Figure 7.13.

of which performs a different function: (*a*) the seal, which excludes
moisture from the cable insulation, and (*b*) the gland, which connects
the cable to a conduit-entry box. The seal consists of a brass pot with an
insulated disc to close the mouth. Sleeves insulate the conductor tails.
The gland consists of three brass components: a nut, a compression
ring and a body.

There are three types of seal, each being designed for use depending on
the application of the wiring system (see Chapter 8, 'Sheathed wiring
systems'). Wedge-pot seals are common nowadays.

Terminating MICS cable (for use in temperatures up to 100°C)

1. Cut the cable to the length required and allow for an appropriate length of conductor tails. The cable end should be cut off squarely. If the cable has a PVC oversheath, the PVC should be cut back before stripping the copper sheath.
2. Mark the point to which the copper sheath is to be stripped back to expose conductors.
3. Remove the sheath, using one of the recognised methods. Make sure that after stripping, the cable end is squared off and clean and free from burrs.
4. Clean the conductors thoroughly.
5. Slip gland nut, compression ring and gland body onto cable.
6. Screw on sealing pot (forced thread).
7. Slip disc and insulating sleeve assembly on conductors.
8. Press compound into sealing pot from one side.
9. Fit disc into mouth of pot. Crimp sealing pot with crimping tool and clean off surplus compound.

Conductor identification

This subject is dealt with in Chapter 6, 'Conductors and cables'.

Insulation and protection of joints

IEE Regulations require that every joint in a cable shall be provided with insulation not less effective than the original insulation on the cable. The joint must also be protected against the ingress of moisture and mechanical damage. Insulating materials available include pure rubber tape, PVC tape and cambric (Empire tape). Note that black tape is not regarded as a good insulating material and should be used only as an outer protective covering. Depending on the installation conditions in which the jointed or terminated cable is to be used, precautions should be taken to ensure that no damage can occur due to corrosion, high temperatures and the ingress of moisture and water.

Testing

All joints and terminations should be tested electrically as follows:

1. Insulation resistance between conductors.
2. Insulation resistance between each conductor and earth (particularly in metal-sheathed cables).
3. Resistance across termination or joint connection.

8

Wiring systems 1:
Sheathed

A wiring system is an assembly of parts used in the formation of one or more electric circuits. It consists of the conductor, together with its insulation, its protection against mechanical damage (sheathing and/or armouring), certain wiring accessories for fixing the system, and joining and terminating the conductors. Wiring accessories are dealt with in general in Chapter 11. Methods used in the installation of wiring systems and their associated accessories are dealt with in Chapter 12.

As implied by the term 'sheathed wiring system', this method of wiring consists of an insulated conductor provided with a sheath which serves in some degree as a measure of protection against mechanical damage. The insulating materials include impregnated paper, rubber, plastics and mineral insulation. The sheathing materials include lead, tough rubber, plastics, aluminium and textiles. Some of the cables are designed with a view to cheapness and are particularly suited to domestic installations.

TRS (tough-rubber sheathed)
This system is termed an 'all-insulated' system because the sheath is of an insulating material. The insulation is vulcanised-rubber; the sheath is tough rubber. The cable is available in single-core (circular section), flat twin, three-core and twin-with-ECC (earth-continuity conductor). When the system was first introduced it was known as CTS (cab-tyre-sheath), because the outer protective sheathing resembled the type of rubber used in the making of solid tyres for horse-drawn cabs of an earlier age.

The use of the TRS system means that there is no metal associated with it, or at least that the amount of metal used in the installation is the very minimum. This lessens the need for extensive earthing arrangements to be carried out to ensure safety against risk of shock. Among the disadvantages are the weakness of the rubber to withstand severe mechanical damage; and rubber is affected by oil. However, it is cheap and flexible, and easy to instal. Fixing is usually by tinned-brass buckle clips fixed to surfaces with brassed-steel pins. The clips can accommodate various sizes of cables. The. accessories used with the TRS system should, as far as possible, be insulated (generally they are of bakelite). In damp situations, the clips are made of 'ferry' metal and fixed with brass screws. There is a limit to the ambient temperature in which TRS can be used.

PVC (polyvinylchloride sheathed)

Another 'all-insulated' wiring system. The conductors are insulated with PVC and the sheath is of PVC. This plastic material first made its appearance in 1937, from Germany. Although much inferior to rubber, especially in its elastic properties, it is much less affected by oil. The insulation-resistance of PVC is much lower than that of rubber. Generally the accessories available, and the methods of fixing, are much the same as those for TRS. Whereas rubber tends to harden in high temperatures, PVC softens when warmed and this disadvantage precludes it from a number of installation applications. The PVC system is available as single-core (circular), and flat twin, with earth and three-core with earth.

PCP (polychloroprene sheathed)

This form of plastic material was developed because neither TRS nor PVC was quite suitable for farm wiring, where steam, ammonia fumes, lactic acid and milk fats, sulphur fumes, direct sunlight and heat have to be taken into account. The cable consists of conductors insulated with vulcanised-rubber and sheathed with PCP, which is a tough rubber-like substance. The alternative to PCP is a farm-wiring cable which is actually called 'farm-wiring cable'. It consists of TRS cable with an additional protective covering of a partially-embedded braid and compounded overall. Both these cable types are installed in the same way as TRS or PVC. Where junction boxes are used in exposed positions, they should be filled with a plastic insulating compound.

Special cleats are available for the cable where uneven surfaces are encountered, or where spacing away from the fixing surface is needed to avoid a moisture trap.

LAS (lead-alloy sheathed)

The LAS system consists of conductors which are insulated with vulcanised-rubber, taped, and sheathed with a lead alloy (lead with a small proportion of tin and antimony). The lead sheath can be used as an earth-continuity conductor. The sheathing must be continuous throughout the whole installation, either by soldering or by bonding with clamps. Although a certain amount of protection against mechanical damage is provided by the lead sheath, it is, of course, a very soft material and can be damaged severely by blows from any hard object. LAS cables are liable to corrosion where they come into contact with lime in damp situations, or certain types of cement, or with certain hard woods (e.g. oak and beech). The cable is available in single-core (circular), twin-core, three-core and twin-with-ECC. The ECC or bond wire is not always provided; where it is used there is no need to bond the earth sheathing, though the ECC must always be conducted to earthing clamps inside the junction boxes. Fixing is by the same type of buckle clips used for PVC and TRS systems. The system can be adapted for use with a conduit system.

HSOS (house-service overhead system)

The conductors of this wiring system are insulated with vulcanised-rubber and taped. The cores of the cable are then varnished and calico taped; the whole is braided and coated with weather-resisting compound. This type of cable is suitable for consumers' distribution wiring between buildings. They are found either fixed to external walls or supported by a catenary wire between buildings. Ferry-metal buckle clips are used, though porcelain and block-PVC cleats are often used (with brass-screw fixings). The cable is resistant to the adverse effects of all kinds of weather.

PILC (paper-insulated, lead-covered)

Paper-insulated, lead-sheathed and served (compounded jute) cables are mainly used for external underground distribution systems. But this type of cable has a wide application for internal distribution in factories and other industrial premises. It can, therefore, be regarded as a wiring system. The paper is impregnated and must be protected against the ingress of moisture; hence the use of cable sealing-boxes. The current-carrying capacity of this type of cable is greater than that of an equivalent VRI cable. Bending of PILC cables must be done very carefully. Fixing is normally by cleats. Further protection against mechanical damage is provided by armouring in the form of helical-wound steel wire or tape.

MIMS (mineral-insulated, metal-sheathed)

These cables consist of copper (or aluminium) conductors contained in a copper (or aluminium) sheath; the insulant is compressed mineral magnesium oxide. The most common type is the MICS cable, with copper as the main metal for conductors and sheath. The advantages of MICS cables are that they are self-contained and require no further protection (even against high temperatures and fire); they are impervious to water and oil, and immune from condensation. Because the conductor, sheath and insulant are inorganic, the cable is virtually ageless and thus has an indefinite life. Installation is simple, though the ends of the cable must be sealed off against the ingress of moisture by special terminations. Fixing is by clips or saddles. Cables can be obtained with a PVC oversheath. Because of the good heat-resisting properties of the cables, the current rating is higher than that of PVC, TRS or PILC cables. Applications for the cable include industrial installations and hazardous areas. Because the sheath is copper or aluminium, if offers an excellent self-contained ECC. A full range of accessories is available for the system which is adaptable to the screwed-conduit system.

In the past much trouble has been found with MIMS cables as the result of their being used on inductive circuits, such as fluorescent lamps. The trouble arises because, under certain circuit conditions, high voltages appear which can puncture the sheath and cause an earth fault. Any dampness which penetrates to the powder insulation reduces the insulation-resistance and further trouble appears. These types of circuits can be fitted with non-linear resistors, called surge diverters, which act as a shunt to any surge voltage.

MIMS cable can be used for earth-concentric wiring in which the metal sheath acts as an earth-continuity conductor and also as a neutral. The normal 2-core cable can be thus replaced as a single-core cable. Care must be taken with this use of MIMS cable to ensure that bonding is efficient at all boxes and outlets. In the wedge type of seal, the pot has an earth (or neutral) tail welded to the inner part of the wedge.

9

Wiring systems 2:
Conduits, ducts, trunking

This chapter deals with those types of wiring systems based on the provision of preformed sections, used for carrying either separate or integrated cables and conductors. The first of this type was conduit used for carrying wires and to protect them. The idea of iron tubing for carrying water, gas and steam was well known, so it was a natural extension of its use to an electrical purpose. Other materials were used for conduits: zinc, lead and bituminised paper. The real breakthrough for conduit must be credited to L. M. Waterhouse who was born in Edinburgh, and who spent some years as a sea-going electrician. He founded the Simplex Steel Conduit Co. Limited in 1898. The qualities of the system soon proved themselves and led to the development of the system as it is known today. Not only was steel used for conduits, but copper, aluminium, bakelised paper and PVC have been and are still used.

In the very early days of electrical distribution, electric cables were often difficult to obtain. This shortage led to attempts being made to use bare conductors on some form of insulated supports; indeed, to use air as an insulator. Protected copper rods were used in 1873 to supply light to 'Big Ben'; and it was only a matter of time before trunking came onto the market in one form or another. But it was not until recent years, with increased distribution of electrical energy on industrial premises, that the trunking systems came to be accepted. The ducting system (generally called 'underfloor ducting') may be considered as an extension of the conduit system, from which it was actually derived. In the early days, floors were mostly of timber construction. To instal a system was just a matter of lifting floorboards. With changes in building techniques the reinforced concrete floor presented problems of underfloor distribution of electricity, and so the ducting systems came into being to carry cables for the installation circuits.

Steel conduit
The modern steel conduit system is available in two types or classes: Class A, plain-end conduit, and Class B, screwed-end conduit.

Class A conduit

This conduit is known as 'light gauge', plain, slip, pin-grip and lug-grip, according to the types. It has thin walls and is available as close-joint, brazed or welded joint, and solid-drawn. This type of conduit is not heavy enough to withstand threading and so presents problems where earth continuity must be maintained. Various methods for connecting the conduit with the associated accessories are available, including the more acceptable lug-grip, in which the fittings are held together by slipping the conduit end into, say, a box and holding it securely by tightening screws in the lugs. The conduit must be prepared before connecting by removing the enamel. If the contact surfaces are not cleaned, the electrical contact resistance will be too high. The applications of Class A conduit are limited to situations which are not damp and in which the wires do not require a high degree of protection against mechanical damage. Close-joint conduit cannot be bent or set because the seams tend to open. If care is used, slight bends and sets can be made in brazed or solid-drawn conduit. The standard sizes are from 16 mm to 32 mm outside diameter. Fixing is by conduit saddles. The conduits are erected before cables are drawn into them. Unattached terminations are fitted with bush-on rubber or composition bushes to prevent the abrasion of cables.

Class B conduit

This conduit is known as heavy gauge or screwed conduit and is available as seam-welded, and solid-drawn. Alternative finishes are (1) black enamel for internal use in dry situations; (2) silver-grey finish for internal use in dry situations where the conduit is required to match decorations; and (3) hot galvanised or sherardised for external use where the conduit will be subjected to dampness or condensation. Because solid-drawn conduit is more expensive than seam-welded, its use is generally restricted to gas-tight and explosion-proof installation work. Welded-seam conduit is used generally for most good quality installation work. The conduits join with the wide range of associated accessories by means of screw threads, which give good mechanical strength and electrical continuity where the conduit acts as an ECC. The sizes available are from 16 mm to 32 mm, outside diameter. The thread is a shallow electric thread (ET).

A full range of system accessories is available for screwed conduit: bends, elbows, tees and boxes. The first three can be of the inspection type (provided with a detachable lid) or 'solid' (no lid). Boxes have, of course, lids. Inspection bends, elbows and tees of the channel type are permitted, though they are not always suitable. Boxes are preferable to other fittings for drawing in cables, because they allow more space. Rectangular adaptable boxes are used at the intersection of several conduit runs. Fixings include saddles, clips and crampets.

Copper conduit

The advantage of copper conduit is that it resists corrosion and is an excellent ECC. The system is expensive, but has an extremely long life. Conduits are screwed in the normal manner though connections to accessories associated with the system can be soldered. Bronze junction boxes are used. The system has been used in the House of Commons, the Bank of England and in the Rylands Library in Manchester.

Aluminium conduit

This material for conduit did not appear on the market until after the Second World War, when there was an acute shortage of steel. Though in some ways inferior to steel conduit, it has found its own applications because of its light weight, ease of working, and suited to tropical conditions. It is a screwed-end conduit. Precautions must be taken to prevent the occurrence of corrosion where a dissimilar metal is in the vicinity. Because its resistivity is only a little higher than that of copper it offers excellent ECC facilities. The fittings used are generally of cast or pressed aluminium alloy. Galvanised iron or steel fittings can, however, be used. It is not strong mechanically and may require further protection where the risk of mechanical damage is great.

Flexible conduit

Flexible conduits are generally used for the final connections to machinery (e.g. electric motors), where vibration and the possible need to adjust the position to an equipment makes a rigid conduit connection unsatisfactory. Flexible conduit can also deal with the need for complicated bends and sets. It is used for short runs where mechanical damage is unlikely to occur. Flexible conduit made from non-metallic material is dealt with in the following section. Flexible metallic conduit consists of a spirally-wound, partially-interlocked light-gauge galvanised steel strip, and may be watertight or non-watertight. It can be obtained with a PVC oversheath. As the conduit in itself is not a satisfactory ECC, a separate ECC must be run between the special brass adaptors used to join the flexible to ordinary screwed conduit. Sizes available are from 10 mm to 50 mm, internal diameter. Another type of flexible metallic conduit is the 'Kopex' conduit system. This consists of layers of metal and bituminised strip; it is usually supplied in 30 m coils. It is arranged to accommodate standard conduit fittings. It comes in sizes from 12 mm to 75 mm, outside diameter. Its great advantage is that once bent into shape it retains its position: no heating is required. The steel spiral of the conduit does not give satisfactory ECC facilities and so a separate ECC must be run.

Non-metallic conduit

Non-metallic conduits are obtainable in various grades and with the same diameters as steel conduits. There are two main types: (*a*) flexible and (*b*) rigid. The flexible type comes in both round and oval section and is supplied in 7 m coils. The rigid type is supplied in standard lengths. Materials used vary widely; one of the most common is PVC, used with phenolic-moulded fittings which closely resemble the steel-conduit range. Advantages claimed for the non-metallic conduit systems include: elimination of the need for earthing continuity; absence of fire risk due to breakdown in continuity; easy manipulation without the use of special tools; resistance to corrosion from most industrial liquids; no internal condensation takes place.

The flexible type can be bent without tools, but in cold weather needs the application of warmth. The rigid type is bent with the careful application of a flame to soften it. Fixings are by saddles or clips. When required, the conduit will cut its own thread when screwed into a threaded portion. But it can be easily threaded using the normal electric thread stocks and dies. If it is necessary to seal the system, the 'Bostik' type of adhesive may be used. Heavy-gauge PVC can be obtained which will withstand a fair amount of rough treatment both in erection and in service.

Trunking Systems

The following advantages are claimed for trunking:

(*a*) It is much lighter than conduits of the same capacity.

(*b*) Fewer fixings are required for one trunking length than a run of multiple conduits.

(*c*) Wiring is easier and quicker as the cables are 'laid-in' instead of being drawn in.

(*d*) Erection time is reduced.

(*e*) It is an easily adaptable wiring system.

(*f*) Multiple-compartment trunking is available where the segregation of services is required.

Trunking is available in sections (square and rectangular). Lengths are joined by couplers normally secured by screws. Earth straps fixed between each section ensure earth continuity along the trunking run. Trunking is available in both light-gauge and heavy-gauge forms; finish is generally enamel, but a galvanised finish can be supplied for certain installation conditions. There is a very wide range of system fittings which include blank ends, tees, bends of various radii and angles, elbows, couplers, four-way boxes and fire-resisting barriers. Pin-racks are supplied for use in long vertical runs. The cables are wound through them for support.

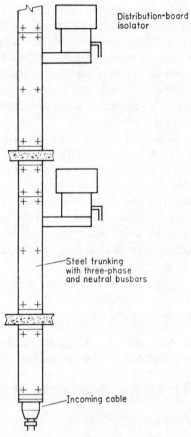

Figure 9.1. Distribution by rising mains.

Where the trunking is used to contain busbars, it becomes (*a*) overhead busbar trunking and (*b*) rising-main trunking. The metal-clad overhead busbar system is used for distribution of electrical energy to machines in factories. The usual arrangement is steel trunking containing copper busbars mounted on insulators. At intervals along the length of run a tapping-off point is provided with three HRC fuses mounted in a sheet-steel case. Three contact blades are designed to fit onto the busbars. Connections to machines are then taken from the tap-off boxes by flexible conduits, steel conduits or other wiring systems. Though this system has a high initial cost, it enables much of the electrical installation work to be carried out before the machines are set in position. Additions and alterations can be carried out quickly.

If long runs of busbars are installed, it is necessary to provide 'expansion' joints at approximately 33 m intervals to allow for expansion and contraction due to changes in temperature. Fire barriers must be installed, particularly where the trunking passes from one room to another (to prevent the spread of fire).

Among other forms of trunking available are:
(*a*) flush trunking, which fits flush with walls; it entails a lot of builders' work to instal; (*b*) multi-compartment trunking, which is of the normal type (square or rectangular) and provided with segregated compartments so that cables carrying different voltages can be accommodated

Figure 9.2. Skirting trunking.

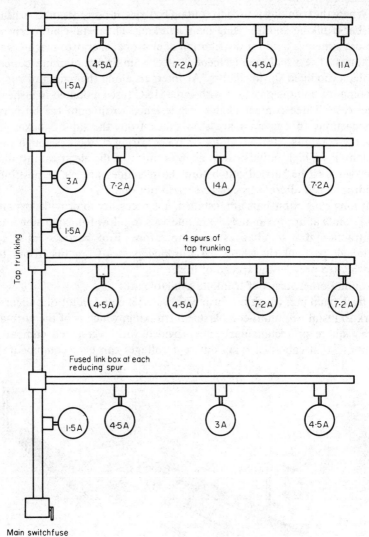

Figure 9.3. Distribution for motors using trunking.

in the same trunking unit; (c) skirting trunking, designed to take the place of the normal room skirting. It carries power, telephone, and lighting cables in its various compartments. Socket-outlets can be easily fitted as an integral part of the trunking; (d) lighting trunking, designed for use where long rows of continuous lighting are required. The steel enclosure not only carries the fluorescent and/or tungsten fittings, but also the control gear and supply cables; (*e*) PVC trunking, with the attendant advantages of this material; (*f*) cable-tap trunking. This type does not carry copper bars, but insulated supports which can accommodate VRI or PVC cables, from which supplies to machines and lighting circuits are tapped through fused tap-off boxes.

Rising mains are used to provide power to the various floors of multi-storey buildings. They are sheet-steel trunking containing copper bars on insulated supports. Provision is made for tapping off at each floor. Where required, distribution boards are fixed direct to the trunking.

Ducting systems

Because a majority of large buildings are now provided with solid concrete floors, the underfloor duct systems were brought into use to solve the problems of providing a distribution network of cables for power and lighting circuits. With the ducting system, the circuits are connected via surface conduits, thence to distribution boards. One of the advantages of this system is that where (e.g. in commercial buildings) a change of tenant is attended by new power layout requirements, the alterations required are easily carried out. Generally, the various ducts are arranged to feed ceiling points for the floor below and socket-outlets for the floor above. Junction boxes are provided with cover plates fixed flush with the floor finish. Because the ducting is laid out in straight lines between junction boxes, it is always possible to locate the runs. Underfloor ducting is available made from sheet metal or an insulated fibre material.

10

Wiring systems 3: Special systems

This chapter deals with those wiring systems which (*a*) have been developed to meet the requirements of specific installation conditions, (*b*) are a combination of the types of wiring systems described in Chapters 8 and 9 and (*c*) are fairly common in use, but are applied only in special circumstances.

Bare conductors

Bare conductors and/or lightly-insulated conductors are installed in buildings for the following purposes only:

1. Earthing connections
2. As the external conductors of earthed-concentric wiring systems
3. Protected rising-main and busbar systems
4. The conductors of extra-low voltage systems
5. As collector wires for travelling cranes, trolleys.

Precautions must always be taken against the possibility of fire risk and earth-shock risk. In an extra-low voltage installation, the insulation must be adequate (light) for the voltage. Protection against fire risk is important because even low voltages can cause sufficient current to flow to overheat conductors and start an electrical fire. In rising-main and busbar systems the conductors, which carry mains voltages of 415V/240V must be inaccessible to unauthorised persons. Strong insulators are needed to support the conductors. The conductors must be able to expand and contract with changes in the ambient temperature and the temperature changes caused by varying load-current conditions. The protection is generally in the form of trunking, or a channel; the metal is earthed. Bare conductors passing through walls, floors, partitions or ceilings must be protected by enclosing them in a non-absorbent incombustible insulating material. Collector wires for travelling cranes and trolleys must be protected by screens or barriers, unless they are so situated that there is no possibility of direct contact. Warning notices indicating the presence of the wires must be fixed along the length of run at intervals not exceeding twelve yards.

Cleated wiring

This is a relatively inexpensive system. It consists of porcelain cleats designed to carry single-core insulated cables (VRI and braided and compounded or PVC-insulated). The cleats have smooth or rounded edges so that the cables carried are prevented from coming into contact with each other or with any other object. If they are installed at a height of 2 m or more, no additional protection against mechanical damage is needed. If protection is needed, a non-metallic conduit is generally used. The cleats are screwed to the fixing surface with wood screws (into wood or fixing plugs). Applications found for this method of wiring include temporary installations, workshops and places where appearance is of no importance.

Catenary systems

The most common system of catenary wiring is that which uses a high-tensile steel wire pulled taut between two fixing positions (e.g. between two buildings). The cable is then run along the catenary wire, being supported in some way. The extension of this type of wiring is in the proprietary wiring system known as the 'Grid-suspension System'. It consists of a central galvanised-steel, high-tensile stranded wire surrounded by a number of PVC-insulated conductors. The whole is enclosed with PVC tape and PVC sheathing. Textile filling is used to obtain a circular section. The catenary wire is secured at each end of its run by eye-bolts and strainers. Special connecting boxes are used at tees and right-angled turns. Applications include the larger types of industrial buildings and similar installations where other types of wiring system would be either difficult or expensive to instal. The system can be adapted to supply feeds for lighting fittings along the length of run.

The system has also been used for overhead street lighting and in factories instead of an overhead busbar system where the amount of power required for machines is small and the expense involved in installing the latter system would be unjustified. The advantages claimed for the system include: completely waterproof; weight imposed on a building is reduced; installation can be left until the building work is completed; reduced installation costs; the system absorbs vibration and lamp-replacement costs are reduced.

Earthed-concentric system

This system is now coming back into use. It consists of a single insulated conductor protected by an earthed metal sheath which is also used as the neutral or return conductor. IEE Regulations B119–124 indicate the specific requirements which must be satisfied before the system can be used. The insulation is PVC and the conductor is a single-core tinned-copper conductor. The system can also be used with single-core and two-core MICS cable.

Wood casing

Though this system is very little used nowadays, it is mentioned because installations once wired with this system are still in existence and the practising electrician may well be called on to carry out repairs on such an installation. The casing consists of a strip of wood with parallel grooves cut lengthways to accommodate VRI cable. The capping is a slat of wood which is screwed to the casing with brass screws. In some tropical countries the system is preferred to conduit because of the problems of excessive condensation associated with the latter.

Overhead system

Though not an actual wiring system, the overhead-cable does provide distribution of electrical energy between points of use, and between buildings where there is some considerable distance between the main control of supply and the point of use (e.g. between a farmhouse and a water-pump a mile away). The cable used in this instance is PBJ. It consists of a conductor of hard-drawn copper, lapped with impregnated-paper tapes, then wound with cotton impregnated with a special compound which has a red-lead base; finally it is weatherproofed. Bare stranded conductors are also used, where heavier amounts of electrical power are required to be distributed; they are more associated with supply-authority overhead distribution than with consumers' distribution. PVC insulation is common nowadays.

11

Wiring accessories

An 'accessory' is defined as 'Any device, other than a lighting fitting, associated with the wiring and current-using appliances of an installation; for example, a switch, a fuse, a plug, a socket-outlet, a lampholder, or a ceiling-rose'. The early history of the origin and development of wiring accessories is indicated elsewhere in this book. This chapter deals with their brief description and the regulations which govern their use and installation.

Switches

Circuit switches are circuit-control devices and are dealt with in detail in Chapter 21. Present-day switches are available in a very wide range, and the practising electrician is referred to the catalogues of the various manufacturers of them. The most common type is the 5A tumbler switch, used invariably to control lighting circuits. A 15A version is also available for controlling heavier circuit currents. For ac circuits the micro-gap switch is common; it is small in size and is suitable where inductive loads (e.g. fluorescent circuits) are to be controlled. So far as contacts are concerned, for dc circuits 'Quick-make-and-break' switches are recommended; for ac circuits 'Slow-make-and-break' switches are used.

All switches should be considered from the aspect of the type of circuit they are intended to control, and the installation condition. Where necessary, suitable enclosures must be provided. In a room containing a fixed bath, for instance, the requirement of the IEE Regulations is that circuit-control switches must be situated out of reach of a person standing in the bath, and preferably either outside the bathroom door, or be of the ceiling-mounted, cord-operated type. All single-pole switches are to be connected in the same conductor throughout the installation, which is the live or phase conductor (IEE Regulation A8).

Lampholders

The first experimental lamps were fitted with wire tails for joining to terminal screws, or were more or less permanently mounted in wooden stands. This method, however, required the services of a mechanic to change the lamp; and it led to the development of the types of lamp-holders we know today, details of which are given in Chapter 23. The first bayonet-cap lampholders were made of brass with the interior part of porcelain. The introduction of plastics mouldings led to the intro-duction and wide use of the all-insulated holders sometimes termed 'shock-proof' lampholders. To minimise the danger of electric shock while replacing lamps, the 'Home Office Skirt' came into being in 1921. Wherever possible, insulated lampholders should be used. Lampholders which are fitted with switches must also be controlled by a fixed switch or socket fixed in the same room. Lampholders must not be used on circuits operated at more than 250V. The outer screwed contact of Edison screw-type lampholders must always be connected to the neutral conductor of the circuit (IEE Regulation C25). The use of lampholder plugs is deprecated. The reason for this is that with these accessories there is no provision for earthing; many accidents (fire and shock) have resulted from the ignorant use of lampholder plugs to supply appliances (e.g. electric irons) which are required to be earthed. Types of lamp-holders include the cord-grip and the backplate (batten). The cord-grip type ensures that the flexible cord is secured in such a manner that the weight placed on it by the lamp and, say, a heavy shade does not over-strain the terminal connections. The shade-carrier ring is a provision for carrying a shade, a shade-carrier or a gallery. The backplate lampholder is a type suitable for screwing to a flat surface; straight and angle types are available. The wiring to backplate lampholders is generally by circuit cables (e.g. 1.00 mm^2); the wiring to lampholders such as the cord-grip type is flexible cord. Care must be taken when wiring the lamp-holder, to see that the strands of the flexible are twisted well together and not allowed to splay. The braiding should be cut back neatly.

Ceiling roses
The ceiling rose is a wiring accessory of china, porcelain, or other insulating material, fitted with terminals intended for connection to a flexible cord carrying a pendant to the wiring of an electrical installation. It is thus a termination point in a wiring system. IEE Regulation C18 states that no ceiling rose may be used in a circuit having a voltage normally exceeding low voltage. Ceiling roses which comply with the requirements of Regulation C19, are not to be connected to the fixed wiring in such a manner that one of its terminals remains live when the

associated switch is OFF, unless that live terminal cannot be touched when the ceiling rose is dismantled to the extent necessary for the replacement of the associated flexible cord. Regulation C21 requires that every ceiling rose shall be provided with an earthing terminal for compliance with the associated Regulation D6. Unless the ceiling rose is designed for multiple pendants, it shall not be used for the attachment of more than one outgoing flexible cord. Ceiling roses are provided with cord-grip arrangements to secure the cord as it carries the weight of the lighting pendant.

Socket-outlets and plugs

The socket-outlet and plug is defined as 'a device consisting of two portions for easily connecting to the supply portable lighting fittings and other current-using appliances'. The socket-outlet is designed to be the fixed member and the plug portion carries two or more metal contacts which connect with corresponding metal contacts in the socket portion. The standard socket-outlets are the 2-pin-and-earth types made in four sizes: 2A, 5A, 15A and 30A. There is also the 13A type which is known as a fused plug, because it carries a small cartridge fuse. Many non-standard types can be found. Because the socket-outlet is virtually the only point in an installation to which a user, usually inexperienced, makes a direct connection to a live supply, it follows that great care must be taken to ensure that terminations are correctly made and that polarity is observed at all times. Shuttered-type socket-outlets are recommended where there is the possibility of misuse, especially by children. When used on dc systems, all socket-outlets must be controlled by a switch in the immediate vicinity; on ac systems no switch is necessary (IEE Regulation A57).

The flexible cords connected to plugs must conform to the recognised method of cable and wire identification so that polarity is true and continuous right through to the appliance. Socket-outlets are not permitted to be installed in bathrooms. Full use of the plug cord-grip must be made so that any strain or pulling will not affect the plug terminations.

Clock-connectors

The clock-connector is an outlet with a small 1A cartridge fuse and is designed to supply an electric clock. The current deemed to be taken is negligible. A switch is not necessary for circuit control.

Electric-shaver sockets

These outlets can be installed in bathrooms because of their double-insulation construction, and also because the voltage they supply is generally 110V through a transformer.

Fused-outlets

The fused-outlet is to be found in conjunction with the ring-main circuit using 13A socket-outlets. They are used for supplying fixed appliances such as water-heaters. They may or may not be switched. The outlet fuse must be rated according to the current rating of the appliance served.

Industrialised-building accessories

With the increasing use of wiring systems specially designed for use in industrialised building, it is inevitable that accessories will take on new forms and functions. An example which will become common in the near future is the 'plug-in pendant'. This is a development to cut time during site installation work, by reducing the amount of work involved in connecting the flexible cord of a plain lighting pendant into a ceiling rose. Instead, the wiring is terminated in a backplate holder and the pendant, which is supplied ready-wired and fitted with a lampholder plug, is simply plugged into the holder. A shroud, moulded to the plug, covers both holder and plug. Another system of wiring developed is a factory-assembled and pre-wired PVC skirting with associated 13A socket-outlets. Connections are made for final subcircuit wiring by means of plug-in connectors.

12

Installation methods

Good workmanship

At the beginning of Part I of the IEE Regulations is written: 'Good workmanship and the use of proper materials are essential for compliance with these Regulations'. This is a requirement with which all practising electricians and others must be familiar. For shoddy and careless work can very easily become a source of real danger to those who are to operate electrical equipment and circuits. The good workman is the man who does not merely produce a good-looking, neat job with his tools. The job must also be the result of the sound and considered application of electrical theory. With the ever-increasing degree of complexity in circuits and apparatus which the electrician of today meets in his work, it is becoming more important than ever that theory is seen to go alongside good practice with tools, whatever the shape or form of the latter. The good workman, then, is familiar with the best methods of installing electrical equipment and circuits, and performs his job with a knowledgeable background of the regulations, both non-legal (e.g. IEE Regulations) and statutory (e.g. Factories Act).

General considerations

Before any electrical installation is begun, some careful thought must be given to the factors or conditions which decide the type of wiring system, its associated accessories, the wiring accessories and the electrical equipment and fittings to be installed. The following are some of the most important points to be considered:

(a) *The type of building*. Whether the installation work is for a permanent building, a temporary building or for the extension to an existing building.

(b) *Flexibility*. Whether the wiring system must be one which will allow it to be easily extended or altered at some time in the future. For instance, a building may have different tenants during its life, all of whom may require the provision of installation arrangements suitable to their needs (e.g. offices, shops, stores, showrooms or dwelling spaces). The more consideration given to this factor in the initial stages of installation planning, the less will be the cost of extension or alteration at a future time.

(c) *Installation conditions.* Whether the installation is likely to be subjected to mechanical damage, moisture, fumes, weather, abnormal or subnormal temperatures, inflammable or explosive dust, gas or vapour.

(d) *Appearance.* Whether the building is such that the electrical installation must be hidden, or its appearance can be allowed. For instance, surface-run conduit is not out of place in a workshop, but it would be in a suite of offices where appearance is of primary importance.

(e) *Durability.* Whether the installation is to last for the time of the life of the building or not. Certain systems have a 'life' in normal installation conditions. The following table shows a rough estimate of how long a system should be allowed to exist before replacement or thorough overhaul:

Wiring system		*Approximate years life in normal circumstances*
PVC cables on cleats		20
Steel-conduit system with PVC cables (concealed work)	conduit	35
	cables	20
Steel-conduit system with PVC cables (surface work)	conduit	30
	cables	20
Steel-trunking system with copper bars		35
PVC cables		20
LAS cables		20
MICS cables		25

(f) *Cost.* Whether the amount of money available for the electrical installation part of a building is restricted, or not. This will then dictate to the installation engineer that he must provide the best job for the smallest amount of money, while still making the installation safe to use. Cost is also related to the time of installation. It may be cheaper to instal an MICS system than a conduit system. Even though the MICS system has a high initial cost, money is saved in labour costs.

(*g*) *Safety*. This is an aspect which is extremely important. Apart from minimising any danger which may arise from factor (*c*) above, safety must also be viewed from the point of view of the type of supply and the earthing arrangements available.

The estimating or installation engineer must consider these main points and other minor ones, before the final choice in favour of one or another wiring system is made. A thorough inspection of the wiring systems, accessories and equipment used in different types of buildings will indicate to the practising electrician the initial planning and considerations involved before the work began.

There are many wiring systems on the market today. Each is designed to perform its duty in specific circumstances. Some are better than others: the cheapest system is not always the best. Nor, on the other hand, is the dearest system. The following notes are an indication of the factors involved in the installation of some of the most common wiring systems available at the present time.

Conduit systems (metal)

There are many metal-conduit systems including heavy-gauge screwed steel conduit, light-gauge lug-grip steel conduit, aluminium conduit, copper conduit, flexible steel conduit. A conduit is a tube designed, so far as electrical work is concerned, to carry electric cables. The most common form of conduit is steel made to BS Specification 31. The early conduit is described in Chapter 2 of this book.

There are two classes of steel conduit: Class 'A' and Class 'B'. The latter is screwed conduit made from heavy-gauge steel. It is by far the most popular system for permanent wiring installations for commercial and industrial premises. Among its advantages are that it gives good protection against mechanical damage, it allows easy rewiring, it minimises fire risks and provides good earth continuity. To erect a good-looking and correct conduit installation requires a certain amount of skill. Among its disadvantages are its cost compared with other systems; it is difficult to instal under wood floors; and it is liable to corrode in acid, alkali and other fumes. Under certain conditions, moisture from condensation may occur inside the conduit. Class 'B' conduit is produced in two forms: solid-drawn, or seam-welded. The finishes are black enamel for dry situations, silver grey for dry situations where the finish will match decorations, and hot-galvanised or sherardised finish for external use, or internal use where there is dampness.

The conduit has many different types of system accessories: boxes, fixings, terminations, all of which are necessary to complete the conduit installation.

Class 'A' conduit is light-gauge and is often called 'slip' conduit. Whereas Class 'B' conduit fittings and accessories are screwed together to complete the installation. Class 'A' conduit is not screwed and the parts are connected by slipping the end of a conduit into a socket. The end is then secured by a lug-grip. This class is limited to 18 mm maximum size, is generally used for installations at 250V or below, and where there is definitely no risk of moisture affecting the cables it carries. The conduit is cut by making a deep groove round the circumference with a file, and then bending it round the knee. This leaves a sharp edge which must be filed. Where the conduit end enters a fitting, the end must be cleaned of enamel to make a good contact between the two surfaces for providing good earth continuity. Slight sets can be made by bending round the knee. Care must be exercised, however, otherwise the conduit may collapse.

Installing class 'B' conduit

There is no restriction to the type of cable that may be drawn into a conduit, except high-voltage types. The cables usually associated with conduits are butyl/silicone insulated, and PVC-insulated. The IEE Regulation B100 states that live and neutral cables of associated circuits must be drawn into the same conduit. The size of conduit to be installed depends on the number of cables to be carried. The number should be such that they can be drawn in quite easily, and in no circumstances should it be greater than that indicated in Tables B5 and B6 of the IEE Regulations. These Tables deserve careful examination.

There are two distinct methods of installing screwed conduit: surface and concealed. The first consideration in the installation of surface conduit is to plan the runs. For instance, where several runs are parallel it is necessary to make sure that there is no crossing at points where directions change. Routes should be chosen to keep the runs as straight as possible. Conduits should be, as far as possible, kept away from gas and water pipes. Places where dampness might occur require special consideration. Though manufactured bends, elbows and inspection tees are available, it should be remembered that where possible, all bends should be made by setting the conduit. Use of inspection fittings of the channel type is not good practice because they do not allow sufficient room for drawing in cables. The better method is to use round boxes.

These look much better and have ample capacity, not only for drawing in cables, but to accommodate a few coils of slack cable (which should always be done at draw-in points).

Inspection boxes must be installed in positions where they will remain accessible during the life of the installation. The conduit system for each circuit should be erected before any cables are drawn in.

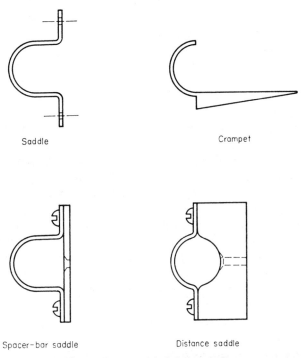

Saddle   Crampet

Spacer-bar saddle   Distance saddle

Figure 12.1. Conduit-fixing methods.

IEE Regulation B89 states that all conduits must be securely fixed. To comply with this requirement there are a number of types of fixings available. Conduit clips are in the form of half a saddle. Though they save using an extra screw (they have only one hole) they are not satisfactory if the conduit is subject to any considerable strain. Saddles are better, and are fixed with screws—not nails. To give the conduit a good fixing, saddles should be spaced at not more than four feet apart, or where required. Where fittings are to be mounted on conduit boxes, the backs of the boxes should be drilled and fixed; otherwise a saddle should be provided on each side of the box and close to it. Spacer-bar saddles are ordinary saddles mounted on a spacing plate with a slot in it. They are used where the conduit leaves conduit fittings. An ordinary saddle would become distorted when screwed down; the spacer-bar saddle takes up the thickness of the fitting and allows the conduit to run straight. Also, the spacer-bar saddle prevents the conduit from making intimate contact with damp plaster, walls and ceilings which might otherwise cause corrosion of the conduit. The distance saddle is used to keep the conduit about 10 mm off the wall, so that any dust which might collect behind conduits can be removed easily. It is for this reason that they are generally specified for places where dust traps must be avoided (e.g. hospitals).

Among other conduit fixings are multiple saddles, where two or more conduits follow the same route, and girder clips, where conduits are run across girders.

When conduit is cut with a hack-saw, burrs are formed on the inner bore of the conduit. If this burr is not removed the insulation of the cables, as they are drawn into the conduit, will be damaged. Burrs are removed with a round file, or with a conduit reamer. If steel conduits are to be installed in damp situations, they must be galvanised or sherardised, and must be Class 'B'. Fixing saddles and screws used must be of the non-rusting type. Suitable precautions must be taken to prevent moisture due to condensation forming inside the conduit. This effect is most likely to occur when the conduit passes from the outside to the inside of a building, where the temperatures are different. In all positions where moisture may collect, holes must be drilled (generally in boxes) at the lowest points in the conduit system so that moisture is drained away. It is generally advisable to use PVC- or TRS-insulated cables, as they are more suitable for moist conditions.

It is an important requirement to ensure that all conduit is not only mechanically sound, but that there is electrical continuity across the conduit joints. The electrical resistance of the conduit, together with the resistance of the earthing lead, measured from the earth electrode, must not exceed twice that of the largest current-carrying conductor of the circuit. So as not to exceed this figure it is necessary to ensure that all conduit connections are tight and that the enamel is removed from places where a metal-to-metal contact is made.

Before drawing any cable into conduit, the system must be fully erected. Also, the conduits must be freed from any obstruction or dampness. This can be done by pulling a small cloth through the conduit (as in cleaning a rifle); this will remove any moisture that may have collected while the conduit has been exposed to weather during building operations. When drawing in cables, the cable reels or drums must be free to revolve, otherwise the cables will spiral off the drums and become twisted, which could damage the insulation. If a number of cables has to be drawn in at the same time, a stand or support should be fixed up so as to permit all the drums to revolve freely. It is good practice to begin the drawing-in operation from a mid-point in the conduit system. This is done to reduce the length of cable that has to be drawn in. A steel tape should be used from one draw-in point to another. The tape is used more to draw in a draw-wire rather than for drawing the cables in. When the draw-wire has been pulled through the conduit by the tape, the ends of the cables should then be attached to the end of the draw-wire. The ends of the cables are bared for about 50 mm and threaded through a loop in the draw-wire. Always when drawing in a number of cables, one man should pull at the receiving end while another feeds the cable in carefully at the other end making sure that no twisting takes place. When necessary, french chalk can be rubbed onto the cable insulation to make the drawing-in operation easier.

Where conduits have to be concealed, they are installed while the building is being erected and can be buried in floors and walls in such a manner that the cables can be drawn into them after the building has been completed, or nearly so. Floors and walls are either solid or hollow, made of cavity-brick, hollow tiles or wood. The conduit runs should be planned so that they run parallel to joist. Where crossing must occur, the joists are slotted to enable the conduit to be kept just below the level of the floorboards. Trap-doors, or short lengths of floorboard must be left at the positions of all junction boxes. These traps must be screwed down and marked.

Where the floors are solid, it is not possible to leave junction boxes in the floors. In this case, the conduit system is arranged so that the cables are drawn in through ceiling or wall points. Conduits to be buried in reinforced concrete should be fixed in position before the concrete is poured. In this instance, special care must be taken to ensure that all joints are tight and painted with a bitumastic paint to prevent rusting. Where the enamel has been removed from the conduit during cutting and threading, paint should be applied.

Installing aluminium conduit

Aluminium conduit is often used as an alternative to steel. It is light in weight, offers good earth continuity and resists corrosion except where it is in contact with other dissimilar metals, and when installed where the atmosphere is polluted with certain chemicals. It is easy to handle, cut and screw, and is non-magnetic. In installing this conduit certain precautions taken in the initial stages will save much time and trouble later on. Corrosion may occur when in contact with other metals, particularly copper, therefore it must be protected by a suitable paint. Conduits in contact with damp cement must be protected by a coating of bitumastic paint. Both cast and pressed aluminium alloy fittings are available for this system: galvanised iron and steel fittings may also be used, but no copper or brass screws, saddles or earth clips. If a copper earth wire is used, the bond should be thoroughly clean and dry and then made moisture-proof by using bituminised paint. Aluminium conduits may be laid directly in the ground if painted with bituminised paint. They must be clean and free from oil and grease. The lubricant to be used when threading aluminium should be a mixture of 50 per cent high-grade mineral oil and paraffin.

Installing copper conduit

The main advantage of copper conduit is that it resists corrosion and provides an excellent earth continuity. The conduit is screwed in the same way as steel conduit, but the screwing of the copper is more difficult and requires care. Connections to conduits of copper can be made by brazing or soldering; bronze junction boxes should be used. Though the system is quite expensive, it is a long-life system, and it is used where freedom from corrosion is of first importance. The House of Commons is provided with a copper conduit system buried in the floors and walls.

Installing flexible metallic conduit

This conduit is made from locked spirals of thin metal and is used for the final connection to motors to provide for the movement of the motor if fixed on slide rails. It also prevents any noise or vibration being transmitted from the motor to other parts of the building. Metallic conduit should preferably be of the watertight pattern and connected by brass adaptors to the steel conduit with which it is used. These adaptors are made so that they screw onto the flexible tubing and also into the steel conduit. Because metallic flexible tubing is not a good earth conductor, an earth wire must be run from the solid conduit system to the frame of the motor. When the tubing is provided with PVC sleeving, it can be used where oil is present.

Conduit systems (non-metallic)

Non-metallic conduits come in both rigid and flexible types. They are supplied in the same diameters as steel conduits. There are a number of advantages claimed for non-metallic conduit systems: there is no need to provide earth continuity; there is an absence of fire risk due to a breakdown in earth continuity; they resist corrosion; no internal condensation takes place. When it is necessary to provide an earth continuity conductor (e.g. at socket-outlet points) a separate wire is drawn into the conduit.

Though this system does not provide great mechanical protection, it does offer many advantages and can be used for specific purposes. Accessories are modelled on the steel conduit types. Rigid conduit can be screwed in the same way as steel conduit. Heavy-gauge rigid types are available for use where unfavourable situations are likely to be encountered. Because they are easy and clean to handle, installation time is a small factor in the total cost of the job. If sealing against the entry of moisture is necessary, Bostik is used. Tools are not needed to bend conduits, though in cold weather they need the application of warmth before they can be worked. IEE Regulation B101 states that all non-metallic conduits shall be non-inflammable, non-hygroscopic, damp-proof, mechanically continuous and strong.

Metal-sheathed wiring systems

The two main metal-sheathed wiring systems in use today are the lead-alloy sheathed (LAS) and the mineral-insulated, metal-sheathed (MIMS).

Cables of the LAS system are insulated with vulcanised-rubber and sheathed overall with a lead-based alloy containing tin and antimony. It is available in circular single-core, or flat 2- and 3-core with or without a copper earth-continuity conductor. The lead sheath is quite flexible and can be 'dressed' easily to fit contours and corners. Though the LAS system was very popular in the past, it is now losing its place of favour. LAS systems may be run on the surface or concealed. Generally they require no further protection, unless they are exposed to mechanical damage. The lead sheath is used as an earth continuity conductor, in which case it is important to ensure that the sheathing is continuous, particularly at the metal junction boxes with which the LAS system is used. At these junction boxes, metal bonding clamps are used. Cables are fixed with clips or saddles. The distances between each fixing varies with vertical or horizontal runs and are indicated in IEE Regulation Table B2. LAS cables must not be allowed to come into contact with extra-low-voltage wiring systems (e.g. bells and telephones), or with gas and water pipes. Cables may be buried in walls without further protection provided there is no danger of nails being driven into the walls to hang pictures and shelves; if there is this danger, then conduit drops must be used. LAS cables are liable to corrode if they come into contact with oak in damp conditions: a coat of bituminised paint will protect the sheathing against attack. A protective sleeving should be provided where the cables pass through floors. Holes must be made good with some incombustible material to prevent the spread of fire. Because lime and certain types of cement can cause corrosion, cables must be protected with light-gauge conduit. All junction boxes used with the system must be accessible throughout the life of the system. Screwed-down traps should be provided. Conductors inside the boxes are connected by porcelain-block conductors or the screw-on or thimble-type porcelain connector. The outside sheathing of the cables must be taken right inside the junction boxes and clamped by the bonding bars to ensure good continuity. Care must be taken when installing LAS cables to avoid kinking: they must be kept flat and not bent at sharp angles; the radius of any bend must be not less than six times the overall diameter of the cable.

Some types of LAS cables have an uninsulated copper conductor immediately under the lead sheathing. On this type of cable there is no need to bond the lead sheath, though the earthing conductor must always be connected to earthing clamps inside the junction boxes. Whether the lead sheath or a bond wire is used as the earth-continuity conductor, the resistance from the earth electrode to any position in the LAS system should not exceed twice that of the largest current-carrying conductor.

LAS cables should be bought on plywood drums, not in unprotected coils, otherwise the sheath may be damaged. Avoid hitting the sheath with a hammer or other sharp object. When 'dressing' the cables, use a smooth wood block; the hammer should strike the wood only. Cable ends are bared by nicking the sheath on each side with a knife and then breaking the sheathing. It can then be withdrawn from the conductors. Always make sure that no sharp edges of the lead sheath project into the rubber insulation of the conductors.

The MIMS system has been in use for a number of decades now, but it is only in the past dozen or so years that it has really become one of the most prominent wiring systems today. The cable consists of copper conductors insulated with a highly compressed magnesium oxide (MgO) powder. The sheath is a seamless copper tube, and this makes the copper-sheathed system (MICS). Where the conductors and sheath are both of aluminium the system is known as the MIAS system. The MICS system is, however, the more popular. Among the advantages claimed for the MIMS system is that the cables are self-contained and need no further protection; the system can withstand high temperatures (250°C, depending on the type of termination) and fire, is impervious to oil and water and is immune to condensation.

Installation of the MICS system is comparatively simple. The cable is generally saddled to walls and ceilings in the same way as LAS cables. Small sizes of the cable can be bent sharply, though it is stated that a bending radius of six times the cable's own diameter will mean that the bend can be straightened out at a later date. Terminations are available for the cable. They must be made with care because the insulation must be kept absolutely dry. Although this type of cable can withstand severe hammering it is necessary to take precautions against the sheath being punctured with possible entry of moisture to the insulation. Conduit is sometimes used with the system to give extra protection against specific types of mechanical damage.

Where the cables pass through floors, ceilings and walls, the holes must be made good with cement to prevent the spread of fire. The sheathing and joint boxes must be bonded throughout the installation to form an earth-continuity conductor. The resistance from the earthing point to any point in the installation using the system must not exceed twice that of the largest current-carrying conductor. The range of glands, reducing nipples and sockets is designed to be accommodated in the standard boxes and fittings of the conduit system, both screw and slip type.

Trunking and ducting
Where a large number of cables have to be installed, or where the cable sizes are large, it is often preferred to use cable trunking rather than conduit. Trunking is rectangular in section, with a cap and is made from sheet steel. There are many variations of the trunking system, which has a full range of fittings and accessories to enable it to meet the specification of any installation. Steel trunking is easy to erect. It can be screwed direct to walls and suspended across trusses. When the run is vertical, the cables should be supported by pin racks. Because the

trunking forms part of the earth-continuity conductor arrangement in the installation, it is necessary, as in the conduit system, to ensure that all sections, fittings and so on of the system are bonded together. This

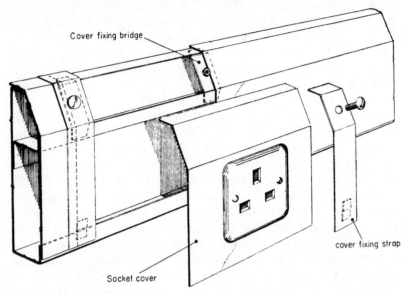

Figure 12.2. Detail of skirting trunking and accessory.

is done either by small copper bars joining each of two parts, or by running a separate conductor in the trunking, bonding this wire to the trunking itself. IEE Regulations B45–52 indicate that cables for lighting and power circuits must not be installed in the same groove or section as cables feeding telephones, bells and ELV alarm systems. However, if the trunking is designed with segregated sections or compartments, then the regulation requirements are satisfied.

When the trunking is designed to contain copper bars, the system is known as busbar trunking, or overhead busbar trunking. The former is used generally for rising mains, supplying mains to each floor of a multi-floor building. The overhead busbar trunking system is used in industrial installations where a considerable degree of flexibility in the electrical provisions is required. Tap-on boxes can be fitted to the trunking at regular intervals throughout the length of the trunking.

The main installation points which apply to trunking are (*a*) it must not be installed where inflammable vapours are present. (*b*) Fire barriers (e.g. asbestos packing) must be installed inside the trunking where it

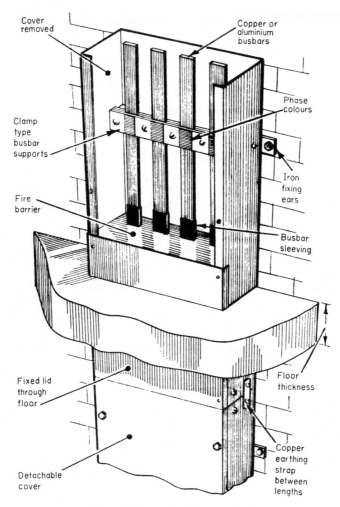

Cover removed

Copper or aluminium busbars

Phase colours

Clamp type busbar supports

Fire barrier

Iron fixing ears

Busbar sleeving

Fixed lid through floor

Floor thickness

Detachable cover

Copper earthing strap between lengths

Figure 12.3. Detail of rising mains.

passes through floors, walls and partitions, and the holes must be made good with cement after the trunking has been installed. (*c*) Allowances should be made in long sections of trunking for expansion in conditions of high temperatures. All busbar trunking should be marked DANGER and the voltage stated. Lids must fit securely.

Ducting is used to provide a network of cable ducts in concrete floors. There are three types of ducting: concrete, steel underfloor and fibre underfloor. Concrete ducts are formed in the ground by erecting shuttering in a channel and pouring concrete mix round the shuttering. This type of ducting is suitable only for armoured cables. The ducts are covered with heavy steel plates. Steel underfloor ducting is very much like heavy-gauge steel trunking with removable outlets. It is placed in position on the floor before the concrete is poured. Fibre ducting is also laid in position before the floor is made.

Cleated wiring system
One of the least expensive methods of wiring is rubber- or PVC-insulated cables supported by cleats made of porcelain or plastic. The system does not give any protection against mechanical damage and so it is useful only for dry situations, and also where the voltage does not exceed 250V. There are a number of points mentioned in the IEE Regulations relating to the installation of cleated wiring.

The cables must be supported on insulated cleats with smooth or rounded edges to prevent damage to the insulation. The insulators must be so spaced to prevent the cables coming into contact either with each other or with another object. The cables must be open to view, and not concealed under floors or partitions or buried in plaster. In damp situations, the supports and fixing must be of a non-rusting material. Where cables are installed at less than six feet from the floor they must be adequately protected. This can be done by using insulating tubing which does not require earthing. Cleats should be fixed in straight-line runs where possible. The cables are then slipped into the cleat grooves and, holding the wires taught, the cleat-screws tightened up. Each run of cleated cable should be anchored at a starting point before pulling the wires taut. Heavier cables should be run in wooden or moulded-insulation material. The runs must be planned beforehand so as to present a good-looking job with no unsightly crossings of wires.

All-insulated wiring systems
This class of wiring system includes TRS (tough-rubber sheathed) and PVC (PVC-sheathed) cables. Their main disadvantage is that they do not offer adequate protection against mechanical damage, though they are relatively cheap and easy to instal. The accessories associated with an all-insulated wiring system are also made from insulating material.

The following points are to be observed when installing all-insulated wiring systems. Wherever possible the cores of the cables must be identified by colours. Note, however, that this cannot apply to a twin cable feeding a switch, where one core, the live, is red in accordance with the Regulations, and the other, the switch wire, is still the live part of the circuit, though coloured black. At switches, socket-outlets and lighting fittings and junction boxes, the sheath of the cable must be taken inside the accessory. Spacings of fixings are regulated by Table B1 in the IEE Regulations. There is no objection to TRS and PVC cables being buried direct in plaster provided care is taken to ensure that there is no likelihood of damage being caused by, say, nails and screws. In such circumstances it is best to use a conduit drop to contain the feed to switches and socket-outlets. If these isolated lengths of conduit are completely concealed they need not be earthed. Otherwise they must be earthed, particularly if they do not terminate in a hardwood or insulated accessory box. Where farms are to be wired with an all-insulated system, the recommended system is PCP (polychloroprene) sheathed. It can withstand steam, sunlight and the many corrosive elements associated with farm buildings. The alternative is known as 'farm-wiring' cable, which is a TRS cable with an additional protective covering of partially-embedded braid and compounded overall.

Catenary systems

These systems are designed to take supplies from one building to another by overhead means, or else for buildings with high ceilings. In its simplest form, the system is a steel wire strained between two points on which insulated cables are carried. For short runs the steel wire carries the cables which are taped to it. The composite catenary cable consists of a high-tensile galvanised steel wire round which are located the PVC cables. Packings of jute or hemp are used to produce a circular section. The alternative insulation for cores is rubber. When installing this type of system it is important to ensure that the steel wire carries the weight and not the conductors themselves. The proprietary system available at present has a full range of accessories and special fittings. If the system is used out of doors it is advisable to fill the connecting boxes with a plastic compound against the entry of moisture. The catenary wire must be securely fixed at each end. The clearance between the cables and the ground must be adequate. A pole may have to be used to act as an intermediate support between the two ends. The cable, where it leaves the catenary should pass through a suitable glazed porcelain lead-in tube. And drip-loops should be provided at entry points so that rainwater is not led into the building by running along the cable.

New systems

In the present climate of increasing industrialised building, it is inevitable that new systems of wiring should come into existence. Just as the early days of electrical installation work were exciting because of the new products and designs that appeared on the market—some to stay and some to fade away—so our times are interesting for the newly-qualified electrician. Prefabricated and prepackaged wiring is not new. But it is now having a very important impact on wiring methods, particularly where repetition jobs are to be carried out (e.g. on municipal housing sites). The systems on the market at present, or being developed, can be divided into four broad classes:

1. Surface systems in the form of skirtings, architraves and covings complete with purpose-made accessories, and supplied with wiring and consumer's main fuse units.
2. Skirtings, architrave and covings to accommodate a standard range of accessories with which wiring is not supplied.
3. Cable 'harness' systems for installing where ducts, conduits or means of access are provided such as in (2) above.
4. Precut and preassembled metal or plastic-sheathed cables for incorporating in the structural members either on site or in the factory.

Some factory prewired systems of skirting and architrave trunking using push-fit connectors enable a house to be completely wired in one day.

13

Current-using apparatus

Electricity has been one of the most important factors in the social progress made by this country over the past half-century. It affects health, education, housing, standards of living and industrial and agricultural progress. Its universal application is seen in the many different types of current-using appliances found in the home, in the office and in the factory making it possible to perform tasks easily, safely and efficiently. This chapter deals with some of the more common current-using apparatuses, their applications and their general installation requirements.

Water-heating

Electric water-heating did not get under way until the early 1930s. When it did begin to make some impact there was a great outcry from the solid fuel and gas interests about inefficiency in the use of electricity for heating water. However, it was shown that raising the temperature of water by means of electricity is an almost 100 per cent efficient method.

The main feature of the electric element used for water-heating is that it is a resistive conductor, insulated and protected from direct contact with the water. The element material is usually of nickel-chromium alloy and wound in the form of a spiral. The insulating material is compressed mineral-oxide powder which can withstand the high temperatures produced by the element. The sheath is of copper or stainless steel. Elements made in this form are able to withstand rough usage. Other types of element consist of spirals of nickel-chromium alloy enclosed in porcelain insulators and mounted on a central rod. Their construction is such that they can be withdrawn for inspection and repair.

Electric kettle. The electric kettle is probably the most common of all electric appliances. Made in a large variety of patterns, the body is either aluminium or chrome-plated copper. The heating element is spiral-wound resistance wire, insulated with magnesium oxide and sheathed. The element, being preformed and self-contained, can be fitted and removed easily from the body by means of a screw thread. A gasket is placed between the element and the body to make a water-tight joint. Loadings vary from 60 to 3000 watts. Connection is made by means of an appliance connector provided with a sliding earth-contact. Protection against damage, due to accidental operation while dry, is provided by a safety device containing a bimetallic strip and small heating element. Sometimes an auto-ejector device is incorporated which pushes the appliance connector out of the kettle to disconnect the element from the supply should the kettle boil dry.

Free-outlet water-heater. This type is sometimes known as a non-pressure type. Generally, it has a cylindrical container with inlet and outlet pipes. The inlet is connected to the water mains through a valve; the outlet is left open. A heating element and a thermostat are located in the bottom

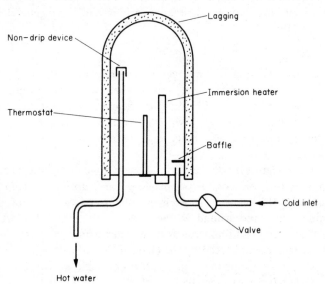

Figure 13.1. Non-pressure-type water heater.

of the container; the latter controls the temperature of the water in the tank. This type of heater is used to give small quantities (6 l to 12 l) of hot water in an instant for washing-up duties. The tank is insulated against heat loss by a lagging of fibreglass or granulated cork.

Pressure-type water-heater. This type has a hot-water tank fed from a

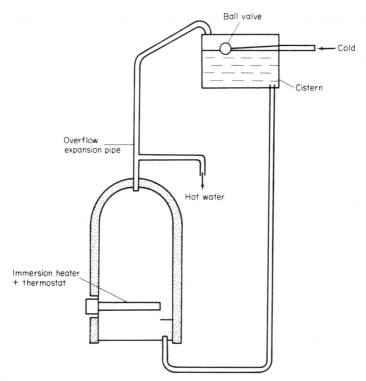

Figure 13.2. Pressure-type water heater.

cold-water cistern placed above it. The water supplied to the cistern is controlled by a ball-valve. Hot-water outlets are thus fed under pressure supplied by the head (vertical height) of the cold water available. This type of water-heater is available in capacities from 5 litres upwards and with loadings from 3kW. Other variations of the pressure-type heaters are the local-storage and the central-storage types; the latter is used in conjunction with a solid-fuel boiler.

Immersion heaters. These are units with self-contained heating elements and thermostats for use where the main heating is provided by a solid-fuel boiler and only supplementary heating is required by electric means. The heater consists of a sheathed-wire element, enclosed in a copper sheath from which it is insulated by compressed magnesium oxide powder. They are generally fixed into a storage water-tank horizontally. The tank must, of course, be efficiently lagged to cut heat losses.

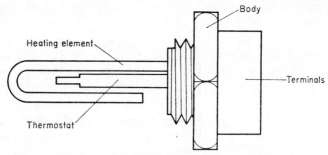

Figure 13.3. Typical water heater, immersion type.

Washboilers. The usual type of washboiler holds about 38 litres of water and has a loading of 3kW. Most types have a three-heat switch and a safety device which cuts off the supply to the heating element should the appliance boil dry.

Installation points

One requirement of the IEE Regulations is that the water must not be in direct contact with that portion of the heating element which is live. The other important requirements are those concerned with (*a*) the current rating of the cables connected to the appliance; (*b*) the earthing arrangements; (*c*) protection against over-loading; and (*d*) the protection of the cables against mechanical damage. These aspects are discussed more fully in other chapters of this book.

Space heating

Electric space-heating falls into three general groups:

1. High-temperature radiation (known as radiant heating).
2. Low-temperature radiation (floor-warming, block-storage and generally 'black-heat' types).
3. Low-temperature convection ('black-heat' used to promote the circulation of warmed air).

Each type of heating has its own advantages and disadvantages and application.

High-temperature radiation

The advantage of direct (radiant) heating is that it can be switched on and off when and where it is required. High-temperature heating has been described as 'personal warming', in that the heat (from a bright-red element) is radiated direct to a specific area: to be warm a person must be in that area.

The radiant fire is probably the most common direct-heating appliance. It is produced in two forms: (*a*) where the element is a spiral of wire fitted in the grooves of a refractory material (fireclay); (*b*) where the element is a wire wound round a long ceramic cylinder and mounted in front of a reflector of polished metal. A variation of type (*b*) is where the heating element is protected by a tubular sheath of metal or silica and mounted in front of an adjustable angle reflector. These are sometimes called 'infra-red' heaters and are particularly suitable for intermittent heating. They are also the type recommended for bathroom and similar situations where there is the presence of moisture or condensation.

Radiant fires are often used to provide background heating, to supplement the general heating from, say, a central-heating system. These radiant heaters all depend on the element becoming red hot. Ratings vary from 750W to 3kW. Heaters can be fixed or portable. Those with two or three separate elements are provided with switches to vary the amount of heat output. All elements should have special guards to prevent direct accidental contact with them (this is a BSI requirement).

Low-temperature radiation

This is an established method of heating. It takes many forms in practice. All are based on a resistive conductor taking an amount of current which will raise its temperature so as to give off radiated heat in small quantities. That is, the element is at 'black heat', which means that the element, though not apparently ON as in the case of the radiator-fire element, still produces heat at a low temperature. Sometimes the elements are embedded between sheets of non-inflammable material such as asbestos, and the sheets fixed to a ceiling. Low-temperature heat is thus radiated downwards into the room. One low-temperature heating system uses heating cables covered by a thermal insulation which forms part of the ceiling structure. Wall-mounted heaters using embedded elements are also used for direct heating; they are often called heater panels. Surface temperatures vary from 49°C to 66°C and have loadings from 80W to 180W per 100 cm^2 of surface. Surfaces with embedded heating cables reach about 38°C.

Low-temperature convection heating

This uses the 'black-heat' principle but instead of producing radiated heat, the air surrounding the element is warmed to create convection currents in the air. Basically the convection heater is constructed to draw cold air in from an inlet at the bottom; the rising air is heated by the element and escapes through an outlet at the top. Convection heaters are used for background heating. One example is the low-temperature tubular heater; air in the vicinity of the heater is warmed and so rises, its place being taken by colder air. This establishes a convection current of warm air which is distributed round the room to raise the overall temperature. This type of heater is often fitted with a thermostat situated in a suitable position in the room.

The following are a few of the appliances and apparatuses used for general heating at the present time:

Under-carpet heaters. This consists of a carpet underfelt with concealed high-tensile electric resistance wire in a plastic sheath. The sheath is usually proofed against any liquid which might be spilled on the carpet and allowed to seep through to the underlay. It is tough enough to withstand continual rough traffic. Carpet underlay is made in the standard carpet sizes and is loaded at about 30W per metre. This is a low-temperature application.

Thermal-storage block-heaters. These are now very common appliances used for central heating in domestic and commercial premises. The usual unit consists of an element surrounded by refractory material, such as concrete. The whole is enclosed in a metal case. Each heater unit is designed to take a continuous charge of electricity over a period of twenty-four hours. The current for these heaters is normally switched on during 'off-peak' periods during the day; the tariff is a cheap-rate type. During the times the current is not on, the heat stored by the heater during its charging periods is released slowly, though some types of unit are fitted with fans to give a boost of heat. Block-storage heaters, as they are sometimes called, are connected to a separate circuit which is separately controlled by a time-switch and has a separate meter.

Floor-warming. This method of heating consists of heating cables buried in the floor (or installed so that they can be removed easily). The temperature produced is about 27°C. Because the flooring is concrete, or some other heat-storing material, the floor-warming systems are a good off-peak heating load. The installation is controlled by a thermostat to maintain a constant temperature in the room.

Liquid-filled radiator heaters. These resemble the familiar pressed-steel water radiators. They are filled with oil, water or a water-based fluid and contain an element usually rated at 3kW. Some are also fitted with thermostats.

Fan convector heaters. These have an electric motor-driven fan placed behind a heating element. The purpose of the fan is to give an increased rate of air circulation. This type of heater has the advantage that, when the element is off, cool air can be circulated in a room during hot weather. The domestic type of this heater is known as a 'turbo-convector'.

Installation points

Portable appliances must be effectively earthed. Flexible cables should be sufficient to carry the load current. And the rating of the fuse protecting the appliance, particularly if fed from a 13A socket-outlet, should be correct. Because the flex is loose, it is subject to more hazards than any other part of an installation. Thus, it is liable to get tugged, rubbed, frayed and stepped on. Flexes should be examined regularly for signs of wear; flex or cord grips should hold the cord sheath securely. Flexibles should be no longer than is necessary to take the supply to an appliance. Appliances are usually supplied with a standard length of flex (generally two metres). If there is no nearby socket-outlet, it is better to provide a new point rather than using an extra-long flex.

User-operated heater appliances

These appliances include electric irons, soldering irons, toasters and similar apparatuses which are used to provide heat for special application.

Electric Iron. This appliance consists of an element in tape form wound on a mica former and insulated with mica. The element is clamped between two steel plates and an asbestos pad. The asbestos pad is used to limit the upward flow of heat from the element. The heat is generally controlled by a bimetallic thermostat fitted with a cam-operated control to obtain temperatures ranging from about 120°C to 204°C. An indicating lamp is sometimes fitted to show when the heating element is ON. It is connected across a resistor wired in series with the element. The voltage drop across the resistor is about 2·5V, sufficient to light an ordinary torch bulb.

Soldering irons. These consist of elements of different types (mica-wound and sheath-type) used to produce heat directed to flow towards a copper bit, the shape of which varies according to the type of duty for which the iron is designed. The wattage of the iron also varies with the type of work to be done.

Installation points

These are similar in nature to those for portable heating appliances.

Motor-operated appliances

In its smallest form, the fractional-kilowatt electric motor has an extremely wide application in domestic, commercial and industrial fields. They include drills, food-mixers, washing machines, spin-driers, refrigerators, ventilating units, hair-driers, fans, clocks and so on. One motor used is the ac series or 'universal' type, so called because it runs on both dc and ac alike. Because ac affects commutation (see Chapter 26: 'Motors and control gear') there is a lot of sparking at the brushes, even on light loads. The application for this type of motor includes vacuum cleaners, portable drills and so on. The size of the ac series motor is usually limited to 75W.

For larger loads, such as washing machines, the motor is one of a number of types of single-phase motors, usually capacitor-start (see Chapter 26).

Fractional kW synchronous motors are used in electric clocks and for other applications where a constant speed is required (the speed depends on the frequency of the supply). This type of motor is self-starting when switched direct-on.

Cooking apparatus

Electric cooking is popular both in the home and in hotels and premises where food has to be prepared. There is an extremely wide range of cooker types, which can be seen on a visit to an electricity showroom. Cooker plates include the solid type, in which the heating element is embedded in a refractory material, and the radiant type (sheathed-element). Each plate can be plugged into a socket for easy installation, removal and replacement. Hot cupboards and grills, eye-level controls, oven lighting, are some of the features of the modern electric cooker. Besides standard cookers there are skillets, rotisseries, table cookers and wall grills. One manufacturer has produced a 'disembodied' cooker system in which the normal parts of a cooker (hotplates, oven, grill and so on) are situated at various points in a kitchen within a specified working area, rather than in one unit as is normal practice.

There are many other applications of electricity which fall into the class of appliance or apparatus and which are to be found in the home, office or for factory light personal duty. These include electric blankets, electric rotary irons, automatic oil and solid-fuel boiler feeds, soil-warming, hedge trimming, electric mowers, and power tools for building operations.

General regulations requirements

One of the most common mistakes in installing appliances and their control switches is that of mounting switches and socket-outlets on wood or some other combustible material without protection. Unless the wiring accessory is enclosed in a box, the cables should terminate in an incombustible enclosure. To meet this requirement a number of types of pattresses are available. Where the temperature is likely to exceed 55°C, then heat-resisting or fireproof cables should be used. This is particularly the case when installing a water immerser in the usual domestic airing cupboard, where ventilation is poor. Yet another point, which is not widely appreciated, is that all fixed heating appliances should be controlled by a double-pole switch.

Conductors feeding appliances should be considered from two aspects: (*a*) the current taken by the appliance and (*b*) the possible voltage drop if the conductor is extra long. An electric cooker should be connected to a separate final subcircuit, the minimum size of cable being 6·0 mm² or its equivalent. The circuit should be protected by a 30A fuse. The wiring should terminate at a cooker control-unit, comprising a double-pole switch with visible ON/OFF indication. If a pilot lamp is fitted inside the unit, this need not be fused locally.

All appliances and equipment used at a voltage exceeding extra-low voltage (Band I) must be protected against the occurrence of dangerous earth-leakage currents. This is often achieved by using 'all-insulated' apparatus or by earthing exposed metal parts as indicated in IEE Regulations Section D. Many British Standards apply to various types of electrical appliances and equipment on the market including, in particular, those appliances termed 'Double-insulated'; these include shavers, hair-driers and foodmixers. No electrical apparatus should be installed where it is exposed to conditions which would damage it. Section C of the IEE Regulations indicates the full requirements concerned with the installation of current-using appliances and apparatus. Regulation E9 indicates that all appliances should be tested periodically for insulation resistance between conductors and, where applicable, between conducting parts and earth.

14

Lighting and power circuits

Series circuits

In the series circuit, the items which use the electric current are connected in series, or one after the other, across the same supply source. The main feature of the series circuit is that the supply voltage is equal to the sum of all the voltage drops across it. These smaller voltages appear across each item in the series circuit. If three lamps, with filaments of equal resistance, are connected in series across a 240V supply, the voltage which will appear across each lamp will be 80V. If the resistances are unequal, the value of the voltage which appears across each lamp will vary according to the resistance of its filament. But the sum of the voltages will still be equal to the voltage of the supply. If 240V lamps are used in such a circuit, each lamp filament will carry current but give little or no light. This is because the voltage across the filament is much less than the rated voltage (240V) of the lamp. If, however, the rated voltage of each lamp is, say, 80V, then full brilliance will be obtained. The series circuit thus has limitations in use for practical purposes. The series circuit also depends on continuity of the conducting path for the completion of the circuit, so that current can flow. If this path is broken, (e.g. a filament breaking) then continuity is lost and the series circuit will not function. Thus, a series lighting circuit in a house would be nothing but an inconvenience; if one lamp went out, a new lamp would have to be found and tried in every lampholder to find out which lamp was the faulty one. This procedure is in fact what must be followed when trying to find which lamp has blown on Christmas tree decoration lights. Also, a loose connection will constitute a break in the series circuit. For all practical purposes, then, the series circuit is not used. Applications of the series *principle* are, however, quite common. Among these are:

1. Christmas tree decoration lights. These are, say 20 amps each rated at 12V and connected in series across a 240V mains supply.
2. Mains droppers, which are resistors connected in series with a current-using device to reduce the voltage supplying the circuit to a value at

which the device will operate. For instance, a 110V motor can be operated from a 240V supply by placing a resistor in the motor circuit which will (*a*) reduce the mains voltage by 130V (to leave 110V) and (*b*) allow sufficient current to pass along the conductors to operate the motor. The disadvantage of series resistors is that consideration must be given to the dissipation of the heat generated. This method is also wasteful of electricity. If ac mains are available, the answer to supplying the motor would be a step-down transformer.

3. Current-limiters, which are on dc circuits (e.g. electric arc lamps) resistors. On ac circuits, however, they are 'chokes' or inductors. They are most commonly found in discharge-lamp circuits (e.g. fluorescent lamps) and they serve to limit the amount of current taken by the electric discharge in the lamp.

4. Three-heat circuits, where two elements are placed in parallel or in series across the supply to give a definite variation in heat output (e.g. cooker grill element).

5. Series-parallel circuits. Often the requirement of a lighting circuit is to be able to switch on lights at will to produce a 'dim' light and a 'bright' light. Use of a series-parallel switch will connect lamps either in series (for 'dim') or in parallel (for 'bright') as required. Applications of this arrangement may be found in hospital wards, and in railway carriage compartments.

Radial lighting circuits
The most common lighting circuit is the parallel arrangement, in which each lamp is connected across the supply to receive the full supply voltage. Also, each lamp can now be controlled individually, a facility not really available with the series circuit. If the filament of a lamp breaks and the lamp goes out, the rest of the circuit will normally remain unaffected. Groups of lamps can also be connected in parallel. Chapter 28 indicates a few circuit examples.

Radial power circuits
Just as the most common circuit arrangement for lamps is the parallel one, so it is for power outlets such as socket-outlets, which are also connected across the supply so that a current-using device when plugged into the socket can receive full mains supply voltage.

Ring-main circuits
The ring-main circuit is supplied with energy from each end, as distinct from the radial circuit which is essentially linear in character, being

supplied at one end only. The main application of the ring-main circuit is (*a*) in supply-mains distribution and (*b*) with 13A flat-pin fused plugs.

In the former application, the supply is fed into a pair of conductors at each end. If the cable breaks at any point, the supply is still available for the consumers. Indication of the break will, of course, be that the consumer at the end of each conductor at the point of the break will receive a reduced voltage due to the voltage drop ($V = IR$ volts) in the cable. The main point is that continuity of supply is available. The application of the ring-main circuit for 13A socket-outlets is very common nowadays. Advantages claimed include more diversity, local protection (the fuse inside each plug rated suitably), and saving in the copper costs of the conductors as compared with installing the equivalent number of socket-outlets of the 5A and 15A types.

Voltage drop in circuits

If any conductor carries a current, there will be a loss of volts between the start and finish of the conductor. This drop is $V = IR$, where V is the number of volts lost or dropped, I is the current passing along the conductor and R is the resistance of the conductor. IEE Regulation B23 indicates the maximum permissible voltage drop for circuits. The drop is measured between the supply terminals (at the supply-intake position) and any, or every, outlet point in the installation (e.g. socket-outlets and lighting points). The effect of voltage drop along a conductor is to reduce the voltage available at an outlet: lamps will not give their full rated light output, and heaters will have their heat output considerably reduced. The following examples are given to show the importance of choosing a cable or conductor to (*a*) carry the load current and (*b*) to satisfy the IEE Regulation B23. Full familiarity with the cable current-rating Tables in the IEE Regulations will make the practising electrician become more appreciative of this aspect of wiring. These examples all take into account the effects of temperature, but not the type of cable sheath; it should be noted that the latter factor is often very important.

Example 1

A 7kW, 240V electric cooker is fed from a 240V, single-phase ac supply position 38 metres away. Find a suitable size of cable (PVC-sheathed, twin-and-earth) to satisfy the load-current and volt-drop requirements. The cable is to be enclosed in conduit and the ambient temperature is 40°C. Coarse excess-current protection is provided.

Maximum permissible volt drop = $2\frac{1}{2}\%$ of 240V = 6V

Load current $= \dfrac{\text{Watts}}{\text{Volts}} = \dfrac{7000}{240} = 29\text{A}$

From Table 3M, Columns 3 and 4, cable size 10 mm^2 is chosen.

Current rating of 10 mm^2 = 40A

Amended current rating with application of rating factor for 40°C and coarse excess-current protection (0·94)

$\qquad = 40\text{A} \times 0\cdot94 = 37\cdot5\text{A}$

Volt drop/Ampere/metre for 10 mm^2 = 4·0 mV

Total volt drop when 38 metres of 10 mm^2 carries 29A

$$= \frac{\text{volt drop} \times \text{load current} \times \text{length}}{1000} = \frac{4 \times 29 \times 38}{1000} = 4\cdot4\text{V}$$

This volt drop is within the required limit; thus 10 mm^2 cable satisfies both load current and volt-drop requirements.

Example 2

A water-heater is rated at 3kW, 240V. It is fed by butyl-rubber insulated, single-core cables enclosed in conduit. The cables operate in an ambient temperature of 50°C and are provided with close excess-current protection. The distance between the heater position and the 240V, single-phase ac distribution board is 28 metres. Choose a suitable cable to satisfy current and volt-drop requirements.

Maximum permissible volt drop = $2\frac{1}{2}\%$ of 240V = 6V

Load current = 12·5A

From Table 7M, Columns 3 and 4, cable size 1·5 mm^2 is chosen

Current rating of 1·5 mm^2 = 17A

Amended current rating by application of a factor of 1·22 (because close excess current protection is provided) and a factor of 0·77 (for an ambient temperature of 50°C = $17 \times 1\cdot22 \times 0\cdot77 = 16\text{A}$

Volt drop/Ampere/metre of 10 mm^2 = 28 mV

Total volt drop when 28 metres of 1·5 mm^2 carries 12·5A

$$= \frac{28 \times 12\cdot5 \times 28}{1000} = 9\cdot8\text{V}$$

As this volt drop exceeds the maximum permissible, a larger cable is chosen: 2·5 mm^2.

Current rating of $2 \cdot 5$ mm^2 = 24A
Amended current rating with application of factors ($2 \cdot 22$ and $0 \cdot 77$)
  = 24 × 1·22 × 0·77 = 27A
Volt drop/Ampere/metre of $2 \cdot 5$ mm^2 = 17 mV
Volt drop when 28 metres carries 12·5A

$$= \frac{17 \times 12 \cdot 5 \times 28}{1000} = 5 \cdot 95\text{V}$$

This figure is within the required limit and so $2 \cdot 5$ mm^2 is a suitable
size of cable for the load.

Example 3

A heating load to a boiler house amounts to 74A. The supply is
240V, single-phase and is 22 metres away. The cable to be used is heavy-
duty MICS, 2-core, with bare sheath but not exposed to touch. The
ambient temperature is 30°C.

Maximum permissible volt drop = $2\frac{1}{2}\%$ of 240V = 6V
From Table 16M, columns 5 and 6, cable size 16 mm^2 is chosen
Current rating of 16 mm^2 = 115A
Volt drop/Ampere/metre of 16 mm^2 = 2·8 mV
The factor for ambient temperature is 1·00 therefore the current
  rating is not amended.
Volt drop when 22 metres of 16 mm^2 cable carries 74A

$$= \frac{2 \cdot 8 \times 74 \times 22}{1000} = 4 \cdot 55\text{V}$$

Thus 16 mm^2 is suitable for this load.

Example 4

An electric motor is rated at 6kW, 240V, single phase, ac. It has an
efficiency of 75 per cent and operates at a power factor of 0·8. The 240V
supply point is 32 metres away. The cable to be used is 2-core, copper
conductors, PVC-insulated and sheathed, clipped direct to a cable tray.
The ambient temperature is 50°C and the class of excess-current protection provided is close.

Maximum permissible volt drop = $2\frac{1}{2}\%$ of 240V = 6V

$$\text{Load current} = \frac{\text{Watts}}{\text{Volts} \times \text{Eff} \times \text{P.F.}} = \frac{6000}{240 \times 0.75 \times 0.8} = 42\text{A}$$

From Table 3M, Columns 7 and 8, cable chosen is 10 mm²

Current rating of 10 mm² = 48A

Amended current rating with application of rating factor of 0·88
  = 48 × 0·8 = 42A

Volt drop/A/m for 10 mm² = 4 mV

Total volt drop when 22 metres of 10 mm² carries load current

$$= \frac{4 \times 42 \times 32}{1000} = 5\cdot35\text{V}$$

Therefore 10 mm² cable would be satisfactory for this load. Note in this case that if the motor is subjected to frequent starting the cable size will have to be increased because of the increased value of average current taken by the repeated starting operations.

Example 5

A portable tool on a building site is required to be operated from a 110V, single-phase ac supply point 24 metres away. The current taken by the tool is 4·7A. Normal ambient temperature applies. The flexible cord is to be 3-core, PVC-sheathed.

Maximum permissible volt drop = $2\frac{1}{2}\%$ of 110V = 2·75V

From Table 22M, Columns 3 and 4, cable size 0·75 mm² is chosen

Current rating of 0·75 mm² = 6A

Volt drop/A/m of 0·75 mm² = 56 mV

$$\text{Volt drop total} = \frac{56 \times 4\cdot7 \times 24}{1000} = 6\cdot3\text{V}$$

As this volt drop is too high, the next size of cable is taken: 1·0 mm²

Current rating of 1 mm² = 10A

Volt drop/A/m of 1·0 mm² = 43 mV

$$\text{Total volt drop} = \frac{43 \times 4\cdot7 \times 24}{1000} = 4\cdot85\text{V}$$

This volt drop is still too high and the next size of cable is taken: 1·5 mm²

Current rating of 1·5 mm^2 = 15A
Volt drop/A/m of 1·5 mm^2 = 31 mV

$$\text{Total volt drop} = \frac{31 \times 4·7 \times 24}{1000} = 3·5\text{V}$$

Again the value is too high, and so 2·5 mm^2 is chosen.

Current rating of 2·5 mm^2 = 20A
Volt drop/A/m = 18 mV

$$\text{Total volt drop} = \frac{18 \times 4·7 \times 24}{1000} = 2·15\text{V}$$

Therefore 2·5 mm^2 cable is suitable for the load.

Note that the above instance has been worked out to show that some problems can take up a great deal of calculation work and the following method is sometimes used for speed:

with
0·75 mm^2 cable, volt drop = 56 × (4·7 × 24) × 10^{-3} = 6·3V
1·00 mm^2 cable, volt drop = 43 × (10·28) × 10^{-3} = 4·85V
1·5 mm^2 cable, volt drop = 31 × 10·28 × 10^{-3} = 3·5V
2·5 mm^2 cable, volt drop = 18 × 10·28 × 10^{-3} = 2·15V

Another method makes use of a close inspection of the columns for current rating and volt drop, and the most likely cable size chosen on trial. The use of a slide rule in these calculations saves much time and reduces the possibility of errors due the bulk of arithmetical work required in these problems.

15

Electrical safety

Electricity is a good servant and an extremely bad master. It takes on the latter aspect when carelessness and ignorance become part and parcel of any type of electrical work, whether carried out by electricians or by householders and 'do-it-yourself' amateurs. The results of ignorance and carelessness, and the misuse of electricity, are injury and death. About 140 people are killed each year through electrical accidents. Injuries of an electrical nature (burning and physical injury resulting from electric shock) are sustained by about 3,000 people each year. These figures underline the fact that adequate knowledge about electricity is a prime requirement of anyone responsible for the installation and maintenance of wiring systems, accessories and current-using equipment. Electricity is also dangerous in that it cannot be seen. Its effects are sometimes visible, but no one can tell if a wire is 'live' or not until it is touched—and then it may be too late.

It is of some significance that more fatal electrical accidents occur in domestic and similar premises than in industrial premises. One detailed survey carried out by a supply authority indicated the following examples of shock sources in reported cases:

Electric cookers	213	Electric fires	40
Washing machines	133	Lighting fittings	21
Socket-outlets	103	Radio and TV sets	17

Thus, it is obvious that the electrician has a great responsibility in ensuring that every electrical installation is safe to use, whether by another experienced electrical person or by a householder. Part 1 of the IEE Regulations lists ten main points taken from various statutory regulations which, if observed, will make for a safe electrical installation. In particular, the IEE Regulations state that 'Good workmanship and the use of proper materials are essential for compliance with these Regulations'.

There are two main causes of electrical accidents. One is carelessness.

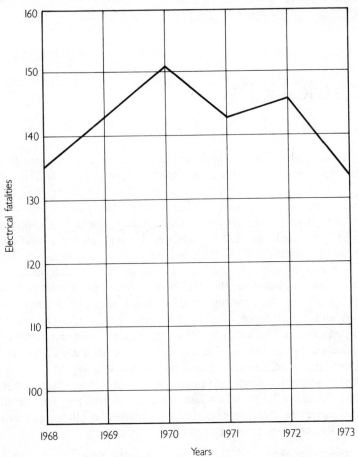

Figure 15.1. Accidents attributed to electrical causes.

This arises when 'familiarity breeds contempt', where an electrician, once having become so used to his work, begins to slacken off his original high standards and attention to small important details. This is very much the case when electricians start to work on or maintain equipment when it is 'live'. This familiarity thus puts danger to the back of the mind—until the last time.

The other cause of electrical accident is ignorance and inexperience. Many electrical tasks seem to be simple, particularly when carried out by an expert. But when they are attempted by some inexperienced person, the job may be left incomplete, with an inherent danger to other persons.

The safety provision in any electrical installation is (*a*) to prevent electric shock and (*b*) to prevent the occurrence of fires due to electrical causes.

Full explanations of these preventive measures are detailed in Chapter 16 'Protection' and in Chapter 17 'Earthing'. However, a few words about the human element in an electrical installation will be useful to give a background knowledge of how electrical accidents can arise.

For an electric shock, it is necessary for the human body to be in contact with two objects of unequal potentials in such a way that the body forms part of an electrical circuit in which current will flow. The

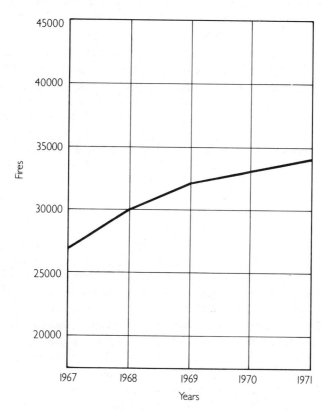

Figure 15.2. Fires attributed to electrical causes.

amount of current flowing through the body will then decide on how serious the accident will become. The currents which cause the following conditions in the human body are:

1–3 milliamps*: This is known as the 'threshold of perception' when a slight tingling sensation is felt.

10–15 milliamps: At this value of current the muscles begin to tighten and it becomes difficult to release any object held (say, a live conductor).

25–30 milliamps: At this current the muscles are really tight and the person has absolutely no control over them. This is the first dangerous state.

over 50 milliamps: At this current fibrillation of the heart occurs, which is generally lethal if immediate specialist attention is not given.

It is thus seen that quite small currents can be fatal, particularly if the person in contact with a live object has a weak heart. As a point of interest, the current taken by a 240V, 15W filament lamp is 62 milliamps.

The most common method used today for the protection of human beings against the risk of electric shock is either (a) the use of insulation (screening live parts, and keeping live parts out of reach) or (b) ensuring, by means of earthing, that any metal in an electrical installation, other than the conductor, is prevented from becoming electrically charged. Earthing basically provides a path of low resistance to earth for any current which results from fault between a live conductor and earthed metal.

The general mass of earth has always been regarded as a means of getting rid of unwanted currents; charges of electricity could be dissipated by conducting them to an electrode driven into the ground. A lightning discharge to earth illustrates this basic concept of the earth as being a large 'drain' for electricity. Thus, every electrical installation which has metalwork associated with it (either the wiring system, accessories or the appliances used) is connected to earth. Basically, this means that if, say, the framework of an electric fire becomes 'live', the resultant current will, if the frame is earthed, flow through the frame, its associated earth-continuity conductor, and thence to the general mass of earth. Earthing metalwork by means of a bonding conductor means that all the metalwork will be at earth potential; or, no difference in potential can exist. And because a current will not flow unless there is a

* A milliamp is one-thousandth of an amp.

difference in potential, then the installation is said to be safe from the risk of electric shock. Reading through Chapter 17 will indicate that the subject of earthing is an important one and merits close attention by anyone who aims to work with electricity.

Effective use of insulation is another method of ensuring that the amount of metalwork in an electrical installation which could become live is reduced to a minimum. The term 'double-insulated' means that not only are the live parts of an appliance insulated, but that the general construction is of some insulating material. A hair-drier and an electric shaver are two items which fall into this category. Also a TRS or PVC wiring system is double-insulated, whereas a VRI/lead-sheathed wiring system is not.

Though the shock risk in every electrical installation is something with which every electrician must concern himself, there is also the increase in the number of fires caused, not only by faults in wiring, but also by defects in appliances. In order to start a fire there must either be sustained heat or an electric spark of some kind. Sustained heating effects are often to be found in overloaded conductors, bad connections, loose-fitting contacts and so on. If the contacts of a switch are really bad, then arcing will occur which could start a fire in some nearby combustible material, such as wood (e.g. the wood block behind a switch). The purpose of a fuse is to cut off the faulty circuit in the event of an excessive current flowing in the circuit. But fuse-protection (as indicated in Chapter 16) is not always a guarantee that the circuit is safe from fire risk. The wrong size of fuse, for instance (e.g. 15A wire instead of 5A wire) will render the circuit dangerous.

Fires can also be caused by an earth-leakage current causing arcing between live metalwork and, say, a gas pipe. Again, fuses are not always of use in the protection of a circuit against the occurrence of fire. Earth-leakage circuit-breakers (see Chapter 17) are often used instead of fuses to detect small fault currents and to isolate the faulty circuit from the supply.

To ensure a high degree of safety from shock-risk and fire-risk, it is thus important that every electrical installation be tested and inspected not only when it is new but at periodic intervals during its working life. Many electrical installations today are anything up to forty years old. And often they have been extended and altered to such an extent that the original safety factors have been reduced to a point where amazement is expressed on why 'the place hasn't gone up in flames before this'. Insulation, used as it is to prevent electricity from appearing where it is not wanted, often deteriorates with age. Old, hard and brittle VR insulation may, of course, give no trouble if left undisturbed and is in a dry situation. But the danger of shock—and fire-risk—is ever present, for the cables may at some time be moved by electricians, plumbers, gas-fitters and builders.

It is a recommendation of the IEE Regulations that every domestic installation be tested at intervals of five years or less. The Completion and Inspection Certificates in the IEE Regulations show the details required in every inspection. And not only should the electrical installation be tested, but all the current-using appliances and apparatus used by the consumer. The values of insulation resistance, earth resistance and so on are indicated in Chapter 19.

The following are some of the points which the inspecting electrician should look for:

1. Flexible cables not secure at plugs.
2. Frayed cables.
3. Cables without mechanical protection.
4. Use of unearthed metalwork.
5. Circuits over-fused.
6. Poor or broken earth connections, and especially signs of corrosion.
7. Unguarded elements of radiant fires.
8. Unauthorised additions to final subcircuits resulting in overloaded circuit cables.
9. Unprotected or unearthed socket-outlets.
10. Appliances with earthing requirements being supplied from 2-pin BC adaptors.
11. Bell-wire (with extra-low voltage insulation) used to carry mains voltages.
12. Use of portable heating appliances in bathrooms.
13. Broken connectors, such as plugs.
14. Signs of heating at socket-outlet contacts.

16

Protection

In electrical work, the term 'protection' is applied to the precautions taken to prevent damage to the various parts of an electrical circuit: wiring system, accessories, fittings, appliances and apparatuses. The prevention of such damage, generally of a physical nature, means the prevention of danger to life, limb and property from shock and fire of an electrical origin. Protection takes many forms, some of which require much more detailed attention than others. This chapter deals with protection against mechanical damage, fire, corrosion and excess currents. Chapter 17, 'Earthing', deals with the methods of protection used to minimise the risk of electric shock in an installation.

Mechanical damage
Mechanical damage is the term used to describe the physical harm sustained by various parts of an electric circuit generally by impact (e.g. hitting a cable with a hammer), by abrasion (e.g. a cable sheath being rubbed against a wall corner) or by collision (e.g. a sharp object falling to cut a cable). IEE Regulations B25–31 deal with the measures which are required to be taken to prevent damage to cables during normal conditions of service. Most cables are provided with a sheath of some kind; the type of sheath, and its material, will indicate the amount of damage it can possibly sustain. Also, conduits, ducts, trunking or casings must be adequately protected where necessary. The armouring found on paper- and plastic-insulated cables is a common form of mechanical protection. So also is the dished lip on the plates of conduit-entry switch boxes to protect the dolly. Protection against mechanical damage is an important factor in the choice of a wiring system or fitting for a particular situation.

Fire risk
Now that many electrical installations are getting old, the fire hazard has increased to a level where fires of an electrical origin are quite

frequent. Electrical fires are caused by (a) a fault, defect or omission in the wiring, (b) faults or defects in appliances and (c) mal-operation or abuse of the electrical circuit (e.g. overloading). The electrical proportion of fire causation today is around the 20 per cent mark. The majority of installation fires are the result of insulation damage, that is, electrical faults accounting for nearly three-quarters of cables and flex fires. Another aspect of protection against the risk of fire is that many installations must be fireproof or flameproof. The definition of a flameproof unit is a device with an enclosure so designed and constructed that it will withstand an internal explosion of the particular gas for which it is certified, and also prevent any spark or flame from that explosion leaking out of the enclosure and igniting the surrounding atmosphere. In general, this protection is effected by wide-machined flanges which damp or otherwise quench the flame in its passage across the metal, but at the same time allows the pressure generated by the explosion to be dissipated.

It is necessary to ensure that all new and overhauled installations contribute as little as possible to an eventual fire through an electrical cause. IEE Regulations B34–38 indicate the requirements to prevent the risk of overheating; and IEE Regulation B40 requires that when installing conductors in any building it is necessary to avoid leaving gaps or holes in walls or floors which may assist the spread of fire. Holes and other openings must be made good with an incombustible material. And where vertical cable ducts are installed, non-ignitable barriers must be fixed at intervals inside the ducts.

It was not until some years after the First World War that it was realised there was a growing need for special measures where electrical energy was used in inflammable situations. Precautions were usually limited to the use of well-glass lighting fittings. Though equipment for use in mines was certified as flameproof, it was not common to find industrial gear designed specially to work with inflammable gases, vapours, solvents and dusts. With progress, based on the results of research and experience, a class of industrial flameproof gear eventually made its appearance and is now accepted for use in all hazardous areas.

There are two types of flameproof apparatus: (a) mining gear, which is used solely with armoured cable or special flexibles; and (b) industrial gear, which may be used with solid-drawn steel conduit, MIMS Cables, aluminium-sheathed cables or armoured cables. Mining gear is known as 'Group I' gear and comes into contact with only one fire hazard: firedamp or methane. Industrial gear, on the other hand, may well be installed in situations where a wide range of explosive gases and liquids are present. Three types of industrial hazards are to be found: explosive gases and vapours—inflammable liquids—and explosive dusts. The first two hazards are covered by what is called 'Group II' and 'Group III' apparatus. Explosive dusts may be of either metallic or organic origin. Of the former, magnesium, aluminium, silicon, zinc and ferro-manganese are hazards which can be minimised by the installation of flameproof apparatus, the flanges of which are well greased before assembly. The appropriate British Standard Code of Practice is CP 1003, *Installation and Maintenance of Flameproof and Intrinsically-safe Electrical Equipment for Industries other than Coal-mining.*

All equipment certified as 'flameproof' carries a small outline of a crown with the letters FLP inside it. The equipment consists of two or more compartments. Each is separated from the other by integral barriers which have insulated studs mounted therein to accommodate the electrical connection. Where weight is of importance, aluminium alloy is permitted. All glassware is of the toughened variety to provide additional strength. The glass is fitted to the apparatus with a special cement. Certain types of gear, such as distribution boards, are provided with their own integral isolating switches, so that the replacement of fuses, maintenance, and so on, cannot be carried out while a circuit is live.

All conduit installations for hazardous areas must be carried out in solid-drawn 'Class B', with certified draw-boxes, and accessories. Couplers are to be of the flameproof type with a minimum length of 50 mm. All screwed joints, whether entering into switchgear, junction boxes or couplers, must be secured with a standard heavy locknut. This is done to ensure a tight and vibration-proof joint which will not slacken during the life of the installation, and thus impair both continuity and flameproofness. The length of the thread on the conduit must be the same as the fitting plus sufficient for the locknut. Because of the exposed threads, running couplers are not recommended. Specially-designed unions are manufactured which are flameproof and are designed to connect two conduits together or for securing conduit to an internally-threaded entry.

Conduits of 20 mm and 25 mm can enter directly into a flameproof enclosure. Where exposed terminals are fitted, conduits above 25 mm must be sealed at the point of entry with compound. Where a conduit installation is subject to condensation, say, where it passes from an atmosphere containing one type of vapour to another, the system must be sectionalised to prevent the propagation of either condensated moisture or gases. Conduit stopper boxes, with two, three or four entries must be used. They have a splayed, plugged filling spout in the cover so that the interior can be completely filled with compound.

When flexible, metal-sheathed or armoured cables are installed, certified cable glands must be used. Where paper-insulated cables are used, or in a situation where sealing is necessary, a cable-sealing box must be used, which has to be filled completely with compound.

The following are among the important installation points to be observed when installing flameproof systems and equipment. Flanges should be greased to prevent rusting. Special care is needed with aluminium-alloy flanges as the metal is ductile and is easily bent out of shape. All external bolts are made from special steel and have shrouded heads to prevent unauthorised interference; bolts of another type should not be fitted as replacements. Though toughened glass is comparatively strong, it will not stand up to very rough treatment; a faulty glass will disintegrate easily when broken. Protective guards must always be in place. Conduit joints should always be painted over with a suitable paint to prevent rusting. Because earthing is of prime importance in a flameproof installation, it is essential to ensure that the resistance of the joints in a conduit installation, or in cable sheaths, is such as to prevent heating or a rise in voltage from the passage of a fault-current. Remember that standard flameproof gear is not necessarily weatherproof, and should be shielded in some way from rain or other excessive moisture.

Being essentially a closed installation, a flameproof conduit system may suffer from condensation. Stopper boxes prevent the passage of moisture from one section to another. Draining of condensate from an installation should be carried out only by an authorised person. Alterations or modifications must never be made to certified flameproof gear. Because flexible metallic tubing is not recognised as flameproof, cables to moveable motors (e.g. on slide-rails) should be of the armoured flexible cable type, with suitable cable-sealing boxes fitted at both ends. It is necessary to ensure that, as far as possible, contact between flameproof apparatus, conduit, or cables, and pipework carrying inflammable liquids should be avoided. If separation is not possible, the two should be effectively bonded together. When maintaining equipment in hazardous areas, care should be taken to ensure that circuits are dead before removing covers to gain access to terminals. Because flexible cables are a potential source of danger, they should be inspected frequently. All the equipment should be inspected and examined for mechanical faults, cracked glasses, deterioration of well-glass cement, slackened conduit joints and corrosion. Electrical tests should be carried out at regular intervals.

Corrosion

Wherever metal is used, there is often the attendant problem of corrosion and its prevention. Electrical installations do not escape this problem and IEE Regulations B41, 42 and Appendix 3 all deal with aspects of corrosion which apply in particular to this problem. There are two necessary conditions for corrosion: (a) a susceptible metal and (b) a corrosive environment. Nearly all of the common metals corrode under most natural conditions. Little or no scientific approach was made to the study of corrosion until the early years of the nineteenth century. Then it was discovered that corrosion was a natural electro-chemical process or reaction by which a metal reverts in the presence of moisture to a more stable form usually of the type in which it is found in nature. It was Humphrey Davy who suggested that protection against corrosion could result if the electrical condition of a metal and its surroundings were changed.

Corrosion is normally caused by the flow of direct electrical currents which may be self-generated or imposed from an external source (e.g. an earth-leakage fault-current). Where direct current flows from a buried or submerged metal structure into the surrounding electrolyte (the sea or soil), no corrosion takes place. It is an interesting fact to record that where a pipe is buried in the soil there is a 'natural' potential of from $-0.3V$ to $-0.6V$ between the pipe and the soil. In electrical installations, precautions against the occurrence of corrosion include:

(a) The prevention of contact between two dissimilar metals (e.g. copper and aluminium).

(b) The prohibition of soldering fluxes which remain acidic or corrosive at the completion of a soldering operation (e.g. cable joint).

(c) The protection of cables, wiring systems and equipment against the corrosive action of water, oil and dampness, unless they are suitably designed to withstand these conditions.

(d) The protection of metal sheaths of cables and metal conduit fittings where they come into contact with lime, cement and plaster and certain hard woods (e.g. oak and beech).

(e) The use of bituminised paints and PVC oversheathing on metallic surfaces liable to corrosion in service.

Under-voltage
This is an electrical protection required by IEE Regulation A64, and is
a provision in the circuit of an electric, motor to prevent automatic
restarting after a stoppage of the motor due either to an excessive drop
in the supply voltage, or a complete failure of the supply, where unexpected restarting of the motor might cause injury to an operator.
These devices are most commonly to be found in dc faceplate starters
and are called under-voltage releases.

Protective enclosures for electrical apparatus
It is often required of a piece of electrical equipment that it be fitted in an
enclosure which will protect it from physical damage because of some
installation condition. Protection against mechanical damage, fire and
corrosion has already been mentioned. The following are some other
types of protection, usually in the form of an enclosure which is
designated as splash-proof, dust-proof, and so on, when so constructed,
protected or treated that satisfactory operation of the enclosed equipment
is not interfered with, if subjected to the specified condition.

(*a*) *Screen-protected.* Enclosure has the openings covered by screens of
wire-mesh, expanded metal or other perforated material.
(*b*) *Totally-enclosed.* Enclosure with no openings for ventilation, but not
necessarily airtight.
(*c*) *Waterproof.* The enclosure will exclude water under prescribed conditions which include a limited period of submersion.
(*d*) *Drip-proof.* Enclosure with openings so protected that liquid or solid
particles falling on it cannot enter to the enclosed apparatus.
(*e*) *Weather-proof.* The enclosure is able to withstand exposure to sun,
rain, mist, snow and airborne particles.

The British Standard 2817, 'Types of Enclosure for Electrical Apparatus', indicates all the forms which enclosures take. All electrical
equipment must be chosen carefully to meet satisfactorily the conditions of the installation—that, is, the enclosure must be appropriate.

Protective relay equipment
The protective relay is basically a series-connected, direct-acting, overload trip coil. It is electromagnetic in operation and consists of either
an electromagnet and armature, or a solenoid and a central plunger. The
coil of the electromagnet (or the solenoid) consists of a few turns of
conductor connected in series with the main circuit. The armature (or
plunger) is arranged to operate the associated circuit-breaker trip

mechanism. The operating current value of such a device is usually adjustable by varying either the magnetic gap of the electromagnet (or solenoid plunger) or a restraining spring. A time-lag or time-delay can be introduced in the relay circuit by using an air or oil dashpot attached to the movement of the trip device, so that a machine or transformer can take an overload for a short time without serious overheating.

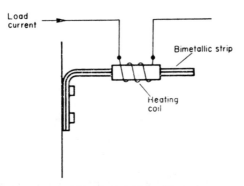

Figure 16.1. Typical bimetallic device.

The thermal relay is suitable for giving overload protection to motors and transformers. The principle of operation is based on the well-known action of bimetallic expansion in which two dissimilar metals, when coupled together in the form of a strip or spring, are arranged to give a deflection under the influence of heat. The amount of deflection increases with the amount of heat, until the point is reached where the operation of the relay results due to the closing of its contacts. If a proportion of the load current is arranged to pass through the bimetal strip or spring, the resultant heat will cause the latter to deflect. By predetermining the amount of deflection required to close the relay contacts, the circuit-breaker can be tripped at any value of the load current it carries. This type of relay has a relatively slow action, because of the time taken to heat up the bimetal element. But this fact is often used as an advantage where overload protection is concerned, since it introduces a suitable time lag. Therefore, it will not operate on momentary or transient overloads such as occur during the starting of a motor, but will do so if the overload is sustained for a predetermined length of time.

Overcurrent
Overcurrent or excess current is the result of either an overload or a short circuit. Overloading occurs when an extra load is taken from the

supply. This load, being connected in parallel with the existing load in a circuit, decreases the overall resistance of the circuit with an attendant rise in the current flowing in the circuit. This increased current will have an immediate effect on the circuit cables: they will begin to heat up. If the overload is sustained the result will be an accelerated deterioration of the cable insulation and the eventual breakdown of it to cause an electrical fault or fire. It is obvious, then, that some means of protection must be incorporated in a circuit to prevent this overloading.

A short circuit is a direct contact or connection between a live conductor and (*a*) a neutral or return conductor or (*b*) earthed metalwork, the contact usually being the result of an accident. The result of the short-circuit is to present a conducting path of extremely low resistance which will allow the passage of a current often of many hundreds of amperes. If the faulty circuit has no overcurrent protection, the cables will heat up rapidly and melt, equipment would also suffer severe damage and fire would be the inevitable result.

Apart from the relays associated with circuit-breakers mentioned already, two methods of overcurrent protection are in wide use: fuses and circuit-breakers. The latter, so far as domestic and small industrial loads are concerned, are miniature circuit-breakers (MCBs).

Fuses

A fuse is defined as 'A device for opening a circuit by means of a conductor designed to melt when an excessive current flows along it. The fuse comprises all the parts of the complete device'. Other terms relating to the fuse are:

(*a*) *Fuse-element*. That part of a fuse which is designed to melt and thus open a circuit.

(*b*) *Cartridge fuse*. A fuse in which the fuse-element is totally enclosed in a cartridge.

(*c*) *Fuse-link*. That part of a fuse which comprises a fuse-element and a cartridge or other container, if any, and either is capable of being attached to fuse-contacts or is fitted with fuse-contacts as an integral part of it.

There are three main types of fuse: the rewirable, the cartridge and the HBC (high breaking capacity) fuse; the latter is a development of the cartridge type.

The rewirable type of fuse consists of a porcelain (usual material) bridge and base. The bridge has two sets of contacts which fit into other contacts in the base. The fuse-element, usually tinned copper wire, is connected between the terminals of the bridge. An asbestos tube or pad is usually fitted to reduce the effects of arcing when the fuse-element melts.

Three terms are used in connection with fuses:

Current rating. This is the maximum current that a fuse will carry indefinitely without undue deterioration of the fuse-element.

Fusing current. This is the minimum current that will 'blow' the fuse.

Fusing factor. This is the ratio of the minimum fusing current to the current rating, namely

$$\text{Fusing factor} = \frac{\text{minimum fusing current}}{\text{current rating}}$$

The rewirable fuse is a simple and relatively cheap type of over-current protective device and is still widely used despite several disadvantages including:

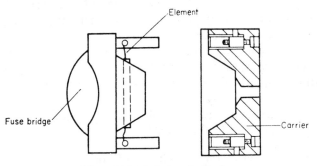

Figure 16.2. Typical rewirable fuse.

(*a*) The ease with which an inexperienced person can replace a 'blown' fuse element with a wire of incorrect gauge or type.

(*b*) Undue deterioration of the fuse-elements due to oxidisation.

(*c*) Lack of discrimination. This means that it is possible, in certain installation conditions, for a 15A fuse-element to melt before a 10A fuse-element. Also, a rewirable fuse is not capable of discriminating between a momentary high current (e.g. motor starting current) and a continuous fault current.

(*d*) Damage, particularly in conditions of severe short circuit.

The fusing factor for a rewirable fuse is about 2. With a protective asbestos pad it is about 1·9. This means that a fuse-element rated at 10A will melt when $10 \times 2 = 20\text{A}$ flows in the circuit. This also means that, say, a $1 \cdot 00 \text{ mm}^2$ conductor (which has a current rating of 11A) may, in an overload condition, be made to carry as much as 50 per cent overload without the fuse coming into action; the cable is thus run on overload which may lead eventually to damage to its insulation.

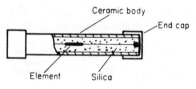

Figure 16.3.

The obvious disadvantages of the rewirable type of fuse led to the development and use of the cartridge fuse which is most often found in 13A fused plugs. Figure 16.3 shows the construction of the cartridge fuse. The fusing factor of this type is about $1 \cdot 5$. Thus, a cartridge fuse rated at 10A will 'blow' at 15A. The rating of this type of fuse is determined and fixed by the manufacturer. Among the advantages of the cartridge fuse are (*a*) the current rating is accurately known and (*b*) the fuse-element is less liable to deterioration in service. Two main disadvantages are (*a*) the fuse-element is more expensive to replace than the rewirable type and (*b*) it is unsuitable for use where extremely high values of fault current may occur.

So far as small domestic and industrial loads are concerned, the following types or ratings are available.

1. House-service Cutout Fuselinks (to BS 88) for use in supply authority cutouts instead of rewirable fuses.
2. Ferrule-cap Fuselinks (to BS 1361) for use in domestic 250V consumer control units, switchfuses, switch splitters, etc. The ratings and identification colours are:

Rating (amps)	Colour code
5	white
15	blue
20	yellow
30	red
45	green
60	purple

3a. Domestic Cartridge Fuselinks (to BS 1363) specifically for use with 13A fused rectangular-pin plugs. The ratings are 3A and 13A.

3b. Domestic Cartridge Fuselinks (to BS 646) specifically for use with 15A, round-pin plugs, where the load taken from the 15A socket-outlet is small (e.g. radio, or a table lamp) in relation to the 15A fuse protecting the subcircuit. The ratings and identification colours are:

Rating (amps)	Colour code
1	green
2	yellow
3	black
5	red

Note that all these cartridge fuses are so designed that they are not interchangeable except within their own group.

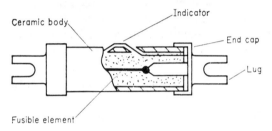

Figure 16.4.

The high breaking capacity fuse (HBC) has its fusing characteristic carefully controlled by the manufacturer. As its name suggests it can safely interrupt very large fault currents. The fuses are often used to protect large industrial loads and main cables. Figure 16.4 shows the construction of a typical HBC fuse. The cartridge barrel is of high-grade ceramic able to withstand the shock conditions when a heavy fault current is interrupted. The end caps and fixing tags are suitably plated to give good electrical contact. The fixing tags are planished when necessary to ensure satisfactory alignment between contact-mating surfaces. Except for very low ratings, the fuse-element is made from pure silver. It has an accurately-machined waist to ensure consistency and reliability. The shape of the waist is designed to give the required operational characteristic.

The filler is powdered silica, carefully dried before use. When used, the filler is compacted in the barrels by mechanical vibration to ensure complete filling. An indicator is provided to show when the fuse has blown. It consists of a glass bead held in position in a recess in the barrel by a fine resistance wire, connected in parallel with the fuse-elements. Barrels are accurately ground and the caps are a force fit. Correct grades of solder are used for the element and tag fixings. The larger types of multi-element fuses have the elements welded in addition to soldering.

The HBC fuse is more expensive than either the rewirable type or the cartridge type. The fusing factor of an HBC fuse is, for small loads, up to 1·25. Thus, a 10A HBC fuse will blow at $10 \times 1·25 = 12·5$A. HBC fuses are discriminating, which means that they are able to distinguish between a high starting current taken by a motor (which lasts only a matter of seconds) and a high fault or overload current (which lasts longer). HBC fuses are often used in motor circuits for 'back-up' protection for the machines. Motors are normally protected against overload by the starter trip; the fuses are required only to give protection against short-circuit currents and overloads outside the capacity of the thermal trip. Modern squirrel-cage induction motors can take up to eight times normal full-load current when stalled. The rating of a fuse-link for a motor circuit should be the smallest current rating that will carry the starting current while providing the necessary margin of safety.

When a capacitor is switched into a circuit, a heavy inrush of current will flow. To ensure that fuses do not blow unnecessarily in these circumstances, it is necessary to fit higher rated fuses. In general, if the fuses fitted are rated at 125 per cent–150 per cent of the capacitor rating, nuisance blowing of the fuses will be avoided. Transformer and fluorescent lighting circuits may also need higher rated fuse-links to deal with the inrush currents associated with this class of gear. Fuselinks with a rating of about 50 per cent greater than the normal current of the apparatus to be protected are usually found to be satisfactory.

One point to remember about fuses and fuse protection is that circuit fuses protect the circuit cables from being overloaded and should also prevent main fuses operating in the case of a local short circuit. Circuit fuses do not protect a current-using device from becoming overloaded, especially when a circuit has more than one outlet.

Circuit-breakers

The circuit-breaker is described in Chapter 21 'Circuit-control devices'. Briefly, it is an automatic device for making and breaking a circuit both under normal and abnormal conditions, such as those of a short circuit. The circuit-breaker has several advantages over any type of fuse:

(*a*) In the event of a fault or overload, all poles are simultaneously disconnected from the supply.
(*b*) It is capable of remote control by means of emergency stop-buttons.
(*c*) Overload and time-lags are capable of adjustment within limits.
(*d*) The circuit can be closed again quickly onto the fault safely.
(*e*) It can open a circuit if the supply fails, thus avoiding unexpected reintroduction of the supply causing apparatus to become live.

The IEE Regulations require that every circuit-breaker shall open a circuit before the current in the circuit exceeds twice that of the rating of the smallest cable it protects (this does not apply to motor circuits). For motor circuits, a time-lag is arranged so that the heavy starting current can be carried for a short period—and satisfactory protection is provided for in normal running conditions (IEE Regulation A68).

The usual circuit-breaker arrangement is one in which an overcurrent is used to open the switch. The switch mechanism is fitted with an arrangement of springs which are compressed as the contacts are closed. Once closed, the mechanism is held in position against the pressure of the springs by a gravity-operated trigger. When released, this trigger causes the mechanism to collapse and open the switch. Solenoid coils are used to carry the maximum current, or some proportion of it, flowing in the controlled circuit. When this current is exceeded the solenoid plunger moves to trip the circuit-breaker.

Recent years have seen the rapid development of the miniature circuit-breakers (MCB) as an alternative to fuses as a means of protection for domestic and small industrial and commercial loads. Among its advantages are:

(*a*) The overcurrent tripping characteristics are set by the maker and cannot be altered.
(*b*) The characteristics are such that the circuit-breaker will trip for a small sustained overload but not for a harmless transient overcurrent (e.g. a filament lamp blowing). Operation is instantaneous when a short-circuit current flows.
(*c*) Faulty circuits are easily identified with the ON and OFF position of the device.

(*d*) The supply can be quickly and easily restored when the fault has been cleared. And if the MCB is switched while the fault is present, it will still be able to 'clear' itself and open satisfactorily.

Though the MCB is now generally accepted as a device for over-current protection, arguments are still put forward against their use, one of the main points being that the circuit-breaker must be tested regularly to ensure that it will perform its protective function when a fault occurs. Cases are on record where an MCB has become sluggish in operation and failed to open, resulting in damage to equipment.

Other types of circuit-breaker include the earth-leakage type (see Chapter 17) which will operate to isolate a fault-to-earth circuit, and the type which includes both overcurrent and earth-leakage protection provisions.

17

Earthing

Earthing terms

Earth. A connection to the general mass of earth by means of an earth electrode.

Earthed. Used to denote an object connected electrically to an earth electrode.

Solidly earthed. Connected electrically to an earth electrode without a fuse, switch, circuit-breaker, resistor or impedance in the earth connection.

Earth electrode. A metal plate, rod, or other conductor buried or driven into the ground and used for earthing metalwork.

Earthing lead. The final conductor by means of which the connection to the earth electrode is made.

Earth-continuity conductor (abbreviated to ECC). The conductor including any clamp connecting to the earthing lead or to each other those parts of an installation which are required to be earthed. The ECC may be in whole or in part the metal conduit, or the metal sheath of cables, or the special continuity conductor of a cable or flexible cord incorporating such a conductor.

Live metalwork. An object is said to be live when:

1. A difference in potential exists between it and earth.
2. It is connected to (*a*) the middle wire, (*b*) common return or (*c*) the neutral of a supply system in which (*a*), (*b*) or (*c*) is NOT permanently and solidly earthed.

Continuous Neutral Earthing (abbreviated CNE). When the CNE system of earthing is used as a means of protection against earth faults in appliances in consumer's premises, the framework of such appliances is connected to the neutral wire of the supply authority's network. And, as an extra precaution, connection is also made to an independent earth of a specified value on the consumer's premises.

Earth resistance. The ohmic resistance between an earth-electrode system and the general mass of earth. In the case of a *consumer's earth*, the resistance is the sum of the resistance of the consumer's ECC and earthing lead, and the resistance of the earth electrode to the general mass of earth.

Earth resistivity. The resistivity in ohms per cubic centimetre (Ω/cm^3) of a sample of earth.

Bond, to. To connect together electrically two or more conductors.

Reasons for earthing

Even before the days of electricity supply on a commercial scale, the soil has been used as a conductor for electrical currents. In early telegraphy systems the earth was used as a return conductor. The early scientists discovered that charges of electricity could be dissipated by connecting a charged body to the general mass of earth by using suitable electrodes, of which the earliest form was a metal plate (the earth plate). But the earth has many failings as a conductor. This is because the resistance of soils varies with their composition. When completely dry, most soils and rocks are non-conductors of electricity. The exceptions to this are, of course, where metallic minerals are present to form conducting paths. Sands, loams and rocks can therefore be regarded as non-conductors; but when water or moisture is present, their resistivity drops to such a low value that they become conductors—though very poor ones. This means that the resistivity of a soil is determined by the quantity of water present in it—and on the resistivity of the water itself. It also means that conduction through the soil is in effect conduction through the water, and so is of an electrolytic nature. Figure 17.1 shows some typical values of resistivity for some soils.

APPROXIMATE VALUES FOR SOIL RESISTIVITY

Description	ohm-cm
Marshy ground	200 to 350
Loam and clay	400 to 15,000
Chalk	6,000 to 40,000
Sand	9,000 to 800,000
Peat	5,000 to 50,000
Sandy gravel	5,000 to 50,000
Rock	100,000 upwards

Figure 17.1. Resistivity of soils.

(*BICC Ltd.*)

For all that the earth is an inefficient conductor, it is widely used in electrical work. There are three main functions of earthing:

1. To maintain the potential of any part of a system at a definite value with respect to earth.
2. To allow current to flow to earth in the event of a fault, so that the protective gear will operate to isolate the faulty circuit.
3. To make sure that, in the event of a fault, apparatus normally 'dead' cannot reach a dangerous potential with respect to earth (earth is normally taken as 0V, 'no volts').

IEE Regulation 4 (in Part I) states that 'where metalwork, other than current-carrying conductors, is liable to become charged with electricity in such a manner as to create a danger if the insulation of a conductor should become defective, or if a defect should occur in any apparatus (i) the metalwork shall be earthed in such a manner as will ensure immediate electrical discharge without danger . . .'. Section D of the IEE Regulations details how Regulation 4 can be complied with. The basic reason for earthing is to prevent or to minimise the risk of shock to human beings. If an earth fault occurs in an installation it means that a live conductor has come into contact with metalwork to cause the metalwork to become live—that is, to reach the same potential or voltage as the live conductor. Any person touching the metalwork, and who is standing on a non-insulating floor, will receive an electric shock as the result of the current flowing through the body to earth. If, however, the metalwork is connected to the general mass of earth through a low-resistance path, the circuit now becomes a parallel-branch circuit with:

(*a*) the human body as one branch with a resistance of, say, 10,000 ohms; and
(*b*) the ECC fault path as the other branch with a resistance of 1 ohm or less.

The result of properly earthed metalwork is that by far the greater proportion of fault-current will flow through the low-resistance path, so limiting the amount of current flowing through the human body. If the current is really heavy (as in a direct short circuit) then a fuse will blow or a protective device will operate. However, an earth fault-current may flow with a value not sufficient to blow a fuse yet more than enough to cause overheating at, say, a loose connection to start a fire.

Regulations

Section D of the IEE Regulations deals with the requirements which all earthing arrangements must satisfy if an electrical installation is to be deemed safe. The main basic requirements are:

1. The complete insulation of all parts of an electrical system. This involves the use of apparatus of 'all-insulated' construction, which means that the insulation which encloses the apparatus is durable and substantially continuous.

2. The use of appliances with double insulation conforming to the British Standard Specifications mentioned in Section F of the IEE Regulations.

3. The earthing of exposed metal parts (there are some exemptions).

4. The isolation of metalwork in such a manner that it is not liable to come into contact with any live parts or with earthed metalwork.

The basic requirements for earthing (D22) are that the earthing arrangements of the consumer's installation are such that the occurrence of a fault of negligible impedance from a phase or non-earthed conductor to adjacent exposed metal, a current corresponding to three times the fuse rating or 1·5 times the setting of an overcurrent circuit-breaker can flow, so that the faulty circuit is made dead. The earthing arrangement should be such that the maximum sustained voltage developed under fault conditions between exposed metal required to be earthed and the consumer's earth terminal should not exceed 40V.

IEE Regulations D2 to D8 detail both the metalwork to be earthed and the exemptions from the requirement. Note that additional precautions are required to be taken near water-pipes (Reg D14) and in bathrooms (Regs D15–D19).

The ECC is the conductor which bonds all metalwork required to be earthed. If it is a separate conductor (insulated and coloured green) it must be at least 1/1·13 (csa = 1·00 mm^2) and need not be greater than 70 mm^2. Note that conduit and trunking may be used as the sole ECC except in agricultural installations (Reg. k29).

Where metal conduit is used as an ECC, a high standard of workmanship in installation is essential. Joints must be really sound. Slackness in the joints may result in deterioration in, and even complete loss of, continuity. Plain slip or pin-grip sockets are not sufficient to ensure satisfactory electrical continuity of joints. In the case of unscrewed conduit, the use of lug-grip fittings is recommended. But for outdoor installations and where otherwise subjected to atmospheric corrosion, screwed conduit should always be used, suitably protected against such corrosion. In screwed conduit installations, the liberal use of locknuts is recommended. Joints in all conduit systems should be painted overall after assembly. In mixed installations (e.g. aluminium-alloy conduit with steel fittings, or steel conduit with aluminium-alloy or zinc-base-alloy fittings) the following are sound recommendations to ensure the electrical continuity of joints.

All threads in aluminium or zinc alloys should be cut using a suitable lubricant. A protective material (e.g. petroleum jelly) should be applied to the threads in all materials when the joints are made up. All joints should be made tight. The use of lock-nuts is advised. In addition, it is recommended to apply bituminised paint to the outside of all joints after assembly. In damp conditions, electrolytic corrosion is liable to occur at contacts between dissimilar metals. To avoid this, all earthing clamps and fittings in contact with aluminium-base-alloy tubing should be of an alloy or finish which is known from experience to be suitable. Copper, or alloys with a high copper content, are particularly liable to cause corrosion when in contact with aluminium-base alloys. For this reason, brass fixing screws or saddles should not be used with conduit or fittings of aluminium-base alloys. Periodical tests should be made to ensure that electrical continuity is satisfactorily maintained. Flexible metal conduits should not be used as an ECC. Where flexible tubing forms part of an earthed metal conduit system, a separate copper or copper-alloy earth-continuity conductor should be installed with the tubing and connected to it at each end.

The earthing lead should be of a minimum size: 6 mm^2, except 2·5 mm^2 is accepted for connection to an earth-leakage circuit-breaker. It must also be protected against mechanical damage and corrosion, and not less than half the largest size of the conductor to be protected, but need not normally exceed 70 mm^2 (D31, D32).

There are a number of methods used to achieve the earthing of an installation:

1. Connection to the metal sheath and armouring of a supply authority's underground supply cable.
2. Connection to the continuous earth wire (CEW) provided by a supply authority where the distribution of energy is by overhead lines.
3. Connection to an earth electrode sunk in the ground for the purpose.

4. Installation of a continuous neutral earthing system.
5. Installation of automatic fault protection.

   One disadvantage in using a mains water-pipe is that sections of the pipe may be replaced by sections of non-conducting material (PVC or asbestos), which makes the pipe an inconsistent earth electrode. IEE Regulation D34 bears close reading. The provision of a cable sheath as an earth electrode connection (method (2)) is very common nowadays. Usually, however, it is accepted that if, for any reason the earthing is subsequently proved ineffective, the supply authority is not to be made responsible. Continuous earth wires are not always provided by the supply authority, except in those areas which have extremely high values of soil resistivity (e.g. peat and rock). The CEW is sometimes called an aerial earth. Connection to an earth electrode sunk in the ground is the most common means of earthing. The earth electrode can be any one of the following forms:

(*a*) *Pipe.* Generally a 200 mm diameter cast-iron pipe, 2 metres long and buried in a coke-filled pit. This type requires a certain amount of excavation; iron is, of course, prone to corrosion, particularly if the coke has a high sulphur content.

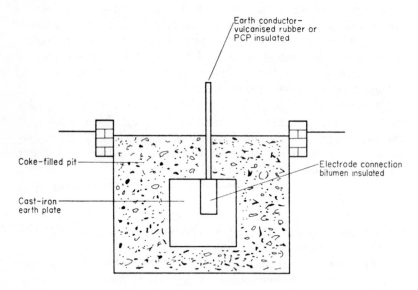

Figure 17.2. Typical earth-electrode pit.

(*b*) *Plate*. Plate electrodes are normally of cast-iron, buried vertically with the centre about one metre below the surface. Copper plates may also be used. Plate electrodes provide a large surface area and are used mainly where the ground is shallow (where the resistivity is low near the surface but increases rapidly with depth). Again, excavation is required. Care is needed to protect the earth-electrode connection (to the earthing lead) from corrosion.

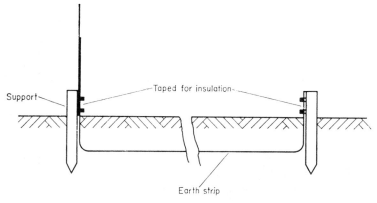

17.3. Earthing using a copper strip.

(*c*) *Strip*. Copper strip is most useful in shallow soil overlying rock. The strip should be buried to a depth of not less than 50 cm.

(*d*) *Rods*. Rod electrodes are very economical and require no excavation for their installation. Because buried length is more important than diameter, the extensible, small-diameter copper rod has many advantages. It can, for instance, be driven into the ground so that the soil contact with the rod is close and definite. Extensible rods are of standard lengths and made from hard-drawn copper. They have a hardened steel tip and a steel driving cap. Sometimes the copper rod has a steel rod running through its centre for strength while it is being driven into rocky soil. Ribbed earth rods have wide vertical ribs to give a high degree of mechanical stiffness, so that they are not easily bent or deflected when driven into the ground.

Because the method used to connect the earthing lead to the earth electrode is important, all clamps and clips must conform to the requirements of IEE Regulation D33 and British Standard 951.

The CNE method (Regulation Appendix 5) gives protection against earth-fault conditions and uses the neutral of the incoming supply as

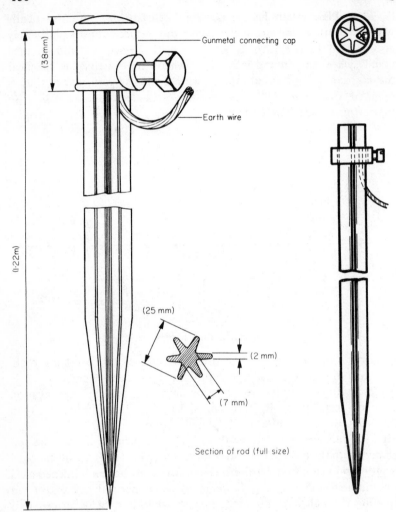

(38mm)

(1·22m)

Gunmetal connecting cap

Earth wire

(25 mm)

(2 mm)

(7 mm)

Section of rod (full size)

Figure 17.4. Ribbed earth-electrode

the earth point or terminal. In this system of earthing, all protected metalwork is connected, by means of the installation ECCs, to the neutral-service conductor at the supply-intake position. By doing this, line-to-earth faults are converted into line-to-neutral faults. The reason for this is to ensure that sufficient current will flow under fault conditions to blow a fuse or trip an overload circuit-breaker, so isolating the faulty circuit from the supply.

The CNE system has a number of disadvantages and stringent requirements are laid down to cover the use of the system (see Appendix 5 in the IEE Regulations). Figure 17.5 shows a typical distribution system with consumers connected to a common CNE system of earthing.

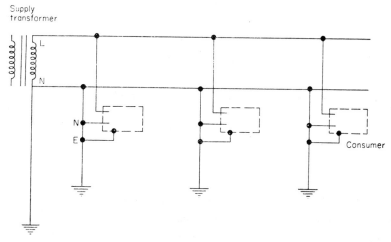

Figure 17.5.

The installation of automatic earth-fault protection equipment has increased in recent years. There are two forms: (*a*) the voltage-operated, earth-leakage circuit-breaker, and (*b*) the current-operated (or current-balance), earth-leakage circuit-breaker.

Voltage-operated units are designed to be directly responsive to fault voltages which appear on protected metalwork, which happens when the earth-path resistance is high (Ohm's Law applies, where $V = IR$; I is the amount of fault-current flowing, and R is the resistance of the earth path). They operate on currents as low as 50mA. To operate correctly, they must be installed with earth electrodes with resistance values up to 500 ohms maximum; which means that with a 500-ohms electrode the unit will operate when the voltage on protected metalwork rises to 40V maximum. The normal operating time of a British Standard (BS842) unit is less than one cycle of ac. Figure 17.6 shows the circuit of a typical voltage-operated ELCB connected to a protected installation. Note that the trip coil is connected between protected metalwork and earth in the same way as one would connect a voltmeter to measure the voltage between the metalwork and earth. If a fault occurs, the current will

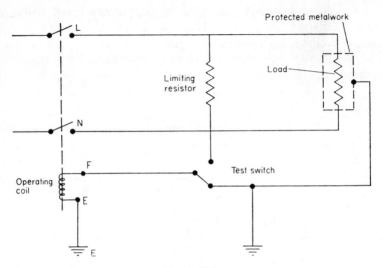

Figure 17.6.

flow through the trip coil to energise it and so trip the breaker contacts. The test switch is required so that the unit can be tested at frequent intervals to make sure that the continuity of the earth path is maintained, and also ensure that the operation of the unit is satisfactory.

Current-operated ELCBs have increased in popularity in recent years.

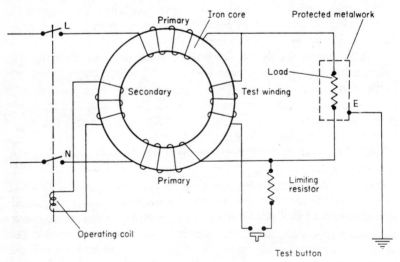

Figure 17.7. Circuit diagram of a current-operated earth-leakage circuit-breaker.

The basic principle of operation depends upon more current flowing into the live side of the primary winding than leaves by the neutral, or other return (earth) conductor. The essential part of the current-operated ELCB is a transformer with opposed windings carrying the incoming and outgoing current. In a healthy circuit, where the values of current in the windings are equal, the magnetic effects cancel each other out. However, a fault will cause an out-of-balance condition and create a magnetic effect in the transformer core which links with the turns of a small secondary winding. An emf (electromotive force measured in volts) is induced in this winding. The secondary winding is permanently connected to the trip coil of the circuit-breaker. The induced emf will cause a current to flow in the trip coil: if this current is of sufficient value the coil will become energised to trip the breaker contacts. A test switch is provided.

Earth-fault-loop path
Figure 17.8 shows the path taken by an earth-fault current. The resistance symbols in both the consumer's earth electrode and the neutral earth electrode indicate the equivalent resistance of the electrode connections.

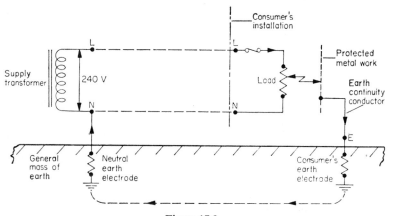

Figure 17.8.

Portable appliances
If the earthing circuit of an appliance becomes completely interrupted, an earth fault on the appliance may cause the casing to become live and constitute a danger to the user. This occurrence is quite common with portable and transportable equipment where the earth conductor

of a flexible cord may break or work loose from its terminal; or no earthing conductor may be provided; or, again, an earthing pin might fail to make effective contact with the earthing socket. In domestic installations, it is always necessary to make frequent checks of such equipment. In industrial installations, the use of voltage-operated ELCB units is generally recommended, with frequent inspections to ensure that the ECC of flexible cords is in good condition to ensure effective operation of the ELCB.

18

Testing and measuring instruments

For a full description of the design, principles of operation and constructional details of the instruments used to measure the current and voltage in an electrical circuit, the reader is referred to the many books on basic electrical science available on the market. This chapter is confined to those instruments, or their variations, in the devices used for testing electrical properties of electrical installations, appliances and apparatus.

Multi-range instruments

These instruments combine a moving-coil movement with a series of shunts and series multipliers to provide a range of readings on a scale graduated to read amperes and volts. Many of them incorporate a small dry-cell battery so that resistance can be measured. The ranges of the instrument are switched from a central position on the instrument case. A rectifier is usually provided so that the movement can be used on ac. There are a number of these multi-range instruments on the market, from the small limited-range instruments to the more complex instruments which are rather expensive, but are designed to read not only current, voltage and resistance, but power factor and decibels (units of sound). The smaller instruments are within the reach of the pocket of mature apprentices and, with correct use and interpretation of their readings, form a most important part of the toolkit.

Testing instruments

While many of the multi-range instruments are used for measuring the electrical properties of materials used in a circuit, they do not necessarily test these materials. For instance, an instrument with a 4·5V battery can hardly be used to test the insulation resistance of a circuit whose cables are to be operated at 240V. Thus, special instruments are needed to simulate the working voltages of circuits to test the insulation and the conductors.

Insulation-resistance tester

To test the value of insulation resistance of the insulating materials used in electrical installations has been a requirement of the electrical engineer since the early days. One of the most popular instruments for this purpose has been the Megger. One type consists of a hand-driven generator which operates a moving-coil instrument. This instrument is similar to the normal mc movement, but has an extra coil fitted on the spindle. One coil is the voltage or pressure coil, and is connected across the generator. The other coil is the current or control coil, which is connected in series with the external resistance (insulation under test). When the external circuit is open (that is, no external resistance connected) only the voltage coil is energised. The field produced by this coil lines up with that of the permanent magnet. And the pointer will move to that end of the scale showing infinity.

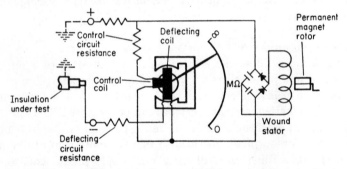

Figure 18.1. Basic circuit diagram of insulation-resistance tester.

When a resistance is connected in the external circuit of the tester, a current will flow through the current coil. The value of this current will depend on the value of the external resistance. The fields produced by both the current and voltage coils interact to produce a movement of the instrument pointer which is a resultant of the two opposing magnetic forces and the field of the permanent magnet. The scale of the instrument is graduated in ohms (e.g. 100 ohms to infinity through a kilohm and Megohm section). The latest models of the insulation-resistance tester have ac generators, the output of which is rectified by static electronic devices instead of a commutator. The voltage output of these testers must, to comply with the IEE Regulations for testing installations, be twice the working voltage to be applied to the installations. The most common tester has an output of 500V dc, though 250V and 1,000V outputs are also available.

Some instruments contain a battery instead of the generator. In these instruments a 9V battery is used in conjunction with a transistor network. The current from the battery passes through the transistor to a step-up transformer. A voltage-doubler rectifier produces the 500V dc test voltage. An adjustable setting resistor compensates for changes in the battery voltage. The movement of the battery tester is a milliammeter movement instead of the two-coil ohmmeter movement. It is in series with the resistance under test.

Continuity tester

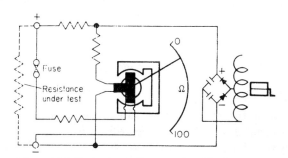

Figure 18.2. Basic circuit diagram of continuity tester.

This instrument is used for testing the electrical continuity of conductors. The readings obtained are in the low-ohms range. They consist of a direct-reading ohmmeter and a small dry battery, usually a 4·5V flash-lamp type. Some instruments combine the function of insulation-resistance and continuity testing. The scales of these instruments are graduated to read both high ohms and low ohms; a switch is used to connect each range. In the battery-operated instrument, the ac convertor is not used, and the operating voltage is that of the 9V battery. The scale for continuity testing is from zero to 100 ohms; that for insulation testing from zero to infinity. The continuity scale is usually open at the low-reading end so that the values up to one ohm can be read easily.

Earth-testing instruments

These instruments are used for testing the resistance of the earth-electrode resistance area. The principle involved in the tests is that a current is passed through the electrode under test and the earth to a distant auxiliary electrode. The potential is measured between the electrode undertest and a central potential electrode. The resistance is obtained by dividing the voltage reading thus obtained by the current flowing in the circuit.

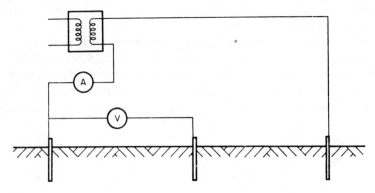

General mass of earth

Figure 18.3.. Circuit diagram for testing soil resistivity.

Earth-loop testers

These instruments are used to measure the resistance of the earth-fault loop. This comprises the line conductor from the point of fault back to the supply transformer, the path through the transformer winding, the earthed neutral point of the transformer, the path from that point to the consumer's earthing lead, the earthing lead itself, and the earth-continuity conductor to the point of fault. Instruments are also available for testing neutral-earth loop impedance.

The use and application of the testing instruments mentioned in this chapter is indicated in Chapter 19.

Ammeters and voltmeters

There are available on the market a large number of instruments for testing voltage and current which do not require to be connected into a circuit. Among these are the very useful 'tong-tester', which has a pivoted jaw and uses the principle of the 'single-turn primary' or 'bar-primary' transformer. The jaw is clamped round a conductor and the scale of the instrument indicates the current in the conductor. Instruments of this type are something of a godsend where a quick check is required. A number can be adapted simply to record voltage by using insulated prods connected into the instrument. Another type of instrument indicates voltage by 'bands', that is a neon lamp will light up to indicate the limits (eg, 50-100 V) within which a certain indicated voltage falls. Particularly where a complex circuit is involved, say in motor starting gear, the circuit can be checked safely while 'live'.

19

Testing and inspection

It is one of the main requirements of the IEE Regulations that every installation shall be tested on its completion and that any defects revealed in the tests or inspection shall be made good. Electrical work is in some ways different from other types of work in that a fault in a circuit may not show itself. It must be revealed either by the use of testing instruments, or by visual inspection. To test and to inspect are two different requirements. The electrical properties of insulation cannot be seen: they can only be measured. And to do this we need instruments. To inspect means to look closely at the electrical work in order to discover if there are any defects which may cause the electrical installation to become a danger to human life. A broken switchbase, for instance, will not show normally as a defect on testing. But the eye will catch sight of it immediately. Again, the continuity of an earthing lead may be extremely good as indicated by a continuity tester, but only a visual inspection will reveal whether or not it is protected or otherwise exposed to mechanical damage.

Section E of the IEE Regulations deals with the testing and inspection of installations. This section follows logically on the natural progress of the previous four sections which indicate generally the sequence or phases of work done to produce a completed installation. Section E indicates that the testing and inspection should be carried out in the following sequence:

1. Verification of polarity.
2. Tests of effectiveness of earthing.
3. Insulation-resistance tests.
4. Test of ring-circuit continuity.

Protective devices and earth-leakage circuit-breakers must also be tried out to make sure that they can perform their protective function when called on to do so. The National Inspection Council for Electrical Installation Contracting (NICEIC) has frequently indicated that there is generally, throughout the country, an insufficient knowledge of the fundamental requirements of testing and inspecting installations. There is also an absence of adequate and/or satisfactory testing instruments and equipment, particularly for testing earth continuity. These two points mean that the practical electrician must not only be able to use testing instruments, but also to interpret their results against a background of good basic electrical theory.

Another point about testing installations: the Electricity Supply Regulations (1937) require that Electricity Supply Undertakings shall not permanently connect an installation unless they are satisfied that the connection would not cause a leakage exceeding one ten-thousandth part of the maximum current taken by the installation. This Regulation can be complied with only if specific tests are carried out. Again, Section 27 of the Regulations stipulates that the Supply Undertakings shall not be compelled to give a supply of energy to any consumer unless they are satisfied that all conductors and apparatus are constructed, installed and protected so as to prevent danger.

Verification of polarity
One of the first tests which must be made concerns a practical point: the need for all single-pole switches (or their variations such as the two-way switch) to be connected in the non-earthed conductor, and for the wiring of plugs and socket-outlets, and the centre contact of Edison-type screw lampholders to be connected in the phase or live pole (see IEE Regulation E2).
Polarity tests are made:

1. With the circuit live. If the circuit is connected to a mains supply a test lamp or voltmeter can be used. There are, however, some disadvantages with this method. For instance, if a 15W lamp is used, it is necessary to ensure that the polarity test is made with all the lamps in the installation removed from their holders. Otherwise the lamp may not light if it happens to be connected in series with a lamp inadvertently left in the circuit under test. With live testing, it is not always possible to distinguish between earth and neutral conductors, though a fully-bright lamp may indeed indicate a good earth connection. When testing live, great care should be taken because of the risk of shock. Only properly-made test lamps should be used, with the leads fitted with insulated test prods.

2. With the circuit dead. If the circuit is not or cannot be connected to a mains supply, a continuity tester or a bell-and-battery set is used. The principle here is to check the continuity of the circuit live con-

ductor from, say, the distribution fuseboard to the switch, lampholder (ES) or socket-outlet concerned. Using an instrument or tester, a zero or almost zero reading should be obtained with correct polarity; the bell will ring to indicate continuity and correct polarity. To summarise, polarity tests ensure that:

(*a*) All single-pole switches or their types are wired in the non-earthed conductor (neutral).

(*b*) The outer or screwed contacts of the Edison-screw type lampholder are connected to the neutral conductor.

(*c*) The live conductor is connected to the terminal marked 'L', the neutral conductor to the terminal marked 'N', and the ECC connected to the terminal marked 'E', in socket-outlets.

Polarity test on SP switch with circuit dead

1. *Before testing*

(*a*) Open main switch and remove the fuses.

(*b*) See that all circuit fuses are wired and inserted in the DFB.

(*c*) Connect the live terminal on the installation side of the main switch to the consumer's main earth terminal.

(*d*) Remove all lamps from their holders and disconnect all appliances.

(*e*) Place all switches to OFF.

2. *Instrument*

Ohmmeter or continuity tester.

3. *Method*

(*a*) Connect one long test-lead of the tester to the consumer's main earth terminal.

(*b*) Connect the other lead of the tester to each terminal in turn of the single-pole switch. Touching one terminal will show a reading of zero or near zero on the tester to indicate correct polarity. Touching the other terminal will give no reading. If no reading is obtained on touching both terminals, the polarity is incorrect, indicating the probable connection of the live conductor to one terminal of the lampholder or ceiling rose.

Variations of this example of polarity-test procedure can be applied to socket-outlets and to ES lampholders, and the switches of appliances. Two-way switches, which have no definite ON position, should be tested as one switch in each pair in both OFF and ON positions. If it is not convenient to use an earth connection, the test can be made using the circuit wiring by testing for continuity between the circuit-fuse start of the live-feed conductor and the switch-feed terminal of the switch.

Insulation-resistance tests

The purpose of the insulation-resistance (IR) tests is to ensure that the quality of the insulating materials used in the installation is good, with particular reference to the circuits' conductors. Thus, the insulation should be such that there is no possibility of leakage currents not only between the conductors, but between the conductors and the general mass of earth. Two IR tests are made on every electrical installation:

(*a*) between each conductor and earth; and
(*b*) between the conductors themselves.

The instrument used for making IR tests is either battery-operated or has a hand-driven generator (see Chapter 18). The testing voltage produced by the instrument should be twice the working voltage of the installation; but it need not exceed 500V for medium-voltage circuits. The voltage must be dc. Because every outlet (i.e. switch, lighting fitting, socket-outlet, etc.) is a potential source of leakage, the IR value to be expected in an installation is inversely proportional to the number of outlets. IEE Regulations E7–E9 indicate the minimum acceptable values of insulation resistance.

Note that the term 'outlet' includes every point and every switch, except that a switch combined with a socket-outlet, appliance or lighting fitting is regarded as one outlet. The minimum acceptable value for any circuit is 1 Megohm. The IR value of an installation will not always remain the same. A good deal depends on moisture and dirt being present on the wiring, fittings and accessories. Deterioration of the insulation is also the result of ageing and unsuitable operating conditions (e.g. rubber in a high ambient temperature).

Insulation-resistance test of conductors to earth

1. *Before testing*

(*a*) Disconnect mains supply by opening main switch and removing main fuses.
(*b*) Make sure that all lamps are in their holders.
(*c*) Place all switches in the ON position.
(*d*) See that all fuses are wired and inserted in the DFB.

(*e*) On incomplete installations, the ends of the conductors at each outlet (points and switch positions) must be bared and joined together.
(*f*) Join together the N and L terminals on the installation side of the main switch.

2. *Instrument*

Insulation-resistance tester (e.g. Megger).

3. *Method*

(*a*) Connect the joined L and N terminals in the main switch to one terminal of the tester.
(*b*) Connect the other lead of the tester to the consumer's main earth terminal.
(*c*) Operate the tester to obtain a reading.

Circuits with two-way switches should be tested twice, with one switch in each pair of strapper wires being changed over before the second test. If the test shows a low reading, withdraw all but one of the circuit fuses in the DFB and test again. Each fuse should be replaced singly to locate the faulty circuit. All readings obtained should be written down.

Insulation-resistance test between circuit conductors

1. *Before testing*

(*a*) Disconnect mains supply by opening main switch and remove main fuses.
(*b*) Remove all lamps from their holders.
(*c*) Place all switches in the ON position.
(*d*) See that all fuses are wired and inserted in the DFB.
(*e*) On incomplete installations, the ends of the conductors at each switch position must be bared and joined together. Those ends of conductors at the lighting points must be kept insulated and not allowed to touch.

2. *Instrument*

Insulation-resistance tester (e.g. Megger).

3. *Method*

(*a*) Connect one lead of the tester to the terminal 'L' in the main switch.
(*b*) Connect the other tester lead to the 'N' terminal in the main switch.
(*c*) Operate the tester to obtain a reading.

Circuits with two-way switches should be tested twice, with one switch in each pair of strapper wires changed over before the second test. It is as well to remember that although faults can occur in straight runs of cables, they are more likely to be found at switches, ceiling roses, lampholders and junction boxes. A zero reading will indicate either a short circuit or else perhaps a lamp not removed from a holder.

Conductor-continuity tests

These tests are made to ensure that conductors are not only continuous throughout their length but that they are truly conductors, that is, that their resistance is zero or nearly zero. The instrument used is the continuity tester or a low-reading ohmmeter. So far as wiring conductors are concerned, the test is to prove that they are not broken, that joints are correctly made and that these joints do not have unduly high resistance. The conductors forming a ring-main circuit should also be tested to ensure that each conductor is in fact continuous throughout its length. The test in this instance should be made from the DFB. The ends of the phase conductor should be separated; the continuity test should be zero or near zero between the two ends. The same test should be made in the neutral and earth conductors.

Earthing tests

The earthing tests prescribed in the IEE Regulations are intended to prove what is called the 'effectiveness of earthing'. This means that whatever arrangement is made to minimise the risk of shock in an installation, it must be tested by one or other of a number of methods. One test is to find the resistance of the earth-continuity conductor (ECC). The IEE Regulation D29 states that where steel conduit or trunking forms part or the whole of the ECC, the resistance should not exceed twice that of the largest current-carrying conductor of the circuit ($\frac{1}{2}$ ohm). If copper or copper-alloy is used for the ECC, the resistance should not exceed 1 ohm. The instrument used for this test can be of the hand-driven generator type (continuity tester) or the battery type (ohmmeter). Because a significant reading is required it is not enough to indicate continuity with a bell-and-battery set. Another method of finding the resistance of the ECC is to pass a current along the ECC and obtain the resistance or impedance (take this as 'ac resistance') value from the ratio R (or Z) $= \frac{V}{I}\,\Omega$, which should not exceed 1 ohm, or $\frac{1}{2}$ ohm depending on the material of the ECC.

To measure the resistance of the ECC

1. *Instrument*

Ohmmeter or continuity tester.

2. *Method*

(*a*) Find the resistance of the instrument leads by shorting their ends together.

(*b*) Measure the resistance of the ECC from consumer's main earth terminal to the farthest point of the ECC in the installation.

Note that the resistance of the instrument leads must be deducted from this reading; also that this test does not take into account the resistance or impedance of the complete earth-fault path of an installation (see Regulation Appendix 6).

A detailed description of the current-injection test is outside the scope of this book. Briefly, however, for background information, the test is carried out to make sure that in the event of a short circuit between a live conductor and exposed earthed metalwork, a current can flow which will either blow a fuse or operate a circuit-breaker. The value of the current injected into the earth-fault path is 1·5 times the rating of the final subcircuit under test, subject to a maximum of 25A (Regulation Appendix 6). There are commercial instruments available on the market for this test, which will make this separate test of the ECC.

Often, however, it is necessary to test the complete earth-fault path. The earth-fault loop as it is called comprises the live conductor from the point of the fault back to the supply transformer, the path through the transformer winding, the earthed neutral point of the transformer, the path from that point to the consumer's earth electrode, the consumer's earthing lead and the ECC to the point of the fault (Regulation Appendix 6). The commercial instruments available test between either line-earth or neutral-earth.

Yet another test is to measure the resistance of that area of ground in which the consumer's earth electrode is installed, called the earth-electrode resistance area test (Regulation Appendix 6). Soils vary greatly in their resistance values (called soil-resistivity values) and so must be tested to ensure that when a fault current flows from an installation into the ground the current is dissipated quickly.

Earth-leakage circuit-breaker trip test

The circuit-breakers must be tested, to ensure that in the event of an earth fault they trip and so isolate the circuit or installation from the supply mains. The circuit-breakers are provided with a test button which the consumer should press at frequent intervals. When the button is pressed, a high-value resistor is connected in series between the live line and the trip-coil which becomes energised to trip the mechanism of the circuit-breaker. The main electrical test is indicated in Regulation Appendix 6.

Appliance and apparatus testing

Although appliances and apparatuses are not strictly speaking part of the electrical installation, they should nevertheless be tested regularly to prove their insulation resistance and any provisions for earth continuity. The value of insulation resistance of an appliance will vary according to its type. However, a reading of not less than 0·5 Megohm should be obtained between conductors and earth. The reading for the resistance of the ECC should be zero or nearly so. The *IR* for fractional horsepower motors should be not less than 1 Megohm. Certain testing instruments are available which pass a current of 15A–20A for a few seconds through the ECC and the metal frame of portable appliances such as drills and other similar tools.

Visual inspection

The visual inspection of an installation is carried out to ensure that the materials, accessories and apparatus and appliances are used in accordance with their designed purpose. IEE Regulations Section F deals with the design and construction of wiring material, accessories and apparatus recommended to be used for electrical installations. Each item should conform to British Standard Specifications and have a guaranteed minimum performance when used in specified working conditions. A visual inspection will reveal that the requirements of Section F are complied with, and also indicate whether or not there are any departures from good practice.

Test certificates

Two test certificates are prescribed in the IEE Regulations. The completion certificate is for the construction of a new installation and certifies on its completion by a competent person that the installation is safe to use. The inspection certificate is a maintenance report on an installation. It is recommended that electrical installations be tested and inspected at intervals of not more than five years. When an installation is completed, the recommended interval should be inserted in the appropriate space. A notice of instruction about the periodic retesting and inspection of an installation should be fixed in a prominent position at or near the main distribution board of every installation (see IEE Regulation E14). Note that the intervals between each inspection and test vary according to the type of installation. For instance, the maximum interval for a temporary installation is 3 months; for a caravan, 1 year; for an agricultural or horticultural installation, 3 years.

20

Temporary installations

The IEE Regulations define a temporary installation as one which is designed to be in service for not more than three months. Temporary installations are most often found on building sites, where electrical services are required during the construction of a building. As might be expected, though the installation is temporary, it is still necessary and just as important to ensure that the safety aspect in the use of electricity on the site is always considered carefully. Section H of the IEE Regulations deals with the requirements which must be satisfied when an installation is classed as 'temporary'. If the installation is required for use for periods longer than three months, then it must be completely overhauled at the end of the three-month period. As soon as the installation is no longer required, it must be disconnected from the supply and dismantled completely.

There are certain types of buildings, and working conditions on sites, which receive the attention of the insurance companies and local byelaws. Thus, not only must the electrician comply with the full requirements of the IEE Regulations regarding temporary installations, but any extra requirements also. For example, though many temporary installations are carried out in some cheap form of wiring system (such as PVC-sheathed), certain types of buildings must be wired in conduit. This is particularly the case where the temporary installation is to supply services for exhibitions in permanent buildings. In many cases the local bye-laws and insurance companies insist on the use of screwed conduit.

Many temporary installations are used to provide lighting and power for a building during the course of its construction. The Regulations permit, in instances where there is adequate protection for the cables, particularly PVC cables. However, if there is a danger of the cables being exposed to mechanical damage or being handled, then protection must be given to the cables in the form of armouring. Metal-sheathed cables, other than mineral-insulated, metal-sheathed (MIMS), must be armoured, and in addition, all of Section D of the IEE Regulations must be satisfied. Where conduit is used to contain the cables, then IEE Regulations B87 to B96 must be satisfied. All cables, and wiring systems, must be adequately supported: a temporary installation does not mean a 'hook-up'.

Where the wiring is exposed to weather conditions, the possibility of danger is increased considerably, particularly to the building operatives handling electrical power tools. To ensure that the danger aspect is minimised or, which is better, completely eliminated, the temporary installation must be in charge of a 'competent person'. This person is generally accepted, particularly by the building industry, as a skilled, fully-qualified electrician.

The electrician is thus fully responsible for the use of the installation and for any alteration or extension. The name and designation of such person shall be prominently displayed close to the main switch or circuit-breaker. This responsibility is given a legal and moral aspect which must be carefully considered by the electrician taking charge of the installation. Only he, or a suitably qualified delegate, is allowed to install new equipment, or make alterations to the installation. The new work must either be supervised and finally checked and inspected by the 'competent person' or by his delegate.

Every temporary installation must be tested and inspected before it is put into service. It must comply, in all respects, to the requirements of the IEE Regulations on insulation-resistance and earth-continuity. In this respect there will be no difference between a temporary installation and a permanent installation.

In the interests of safety, it is always recommended that a low-voltage supply be used. The supply should be through a double-wound transformer which will reduce the supply voltage to 110V the secondary winding being centre-tapped to earth.

A voltage of 110V is now regarded as the standard voltage for supplies on building sites: the use of a reduced voltage on a correctly-earthed supply system, greatly reduces the risk of accident. The bulk of the comprehensive range of power tools available for the various aspects of building work are now rated for 110V. However, though 240V will still be found on building sites, it is always recommended that the lower voltage be used whenever practicable.

The electrician responsible for the distribution of electrical energy on a temporary installation on a building site has many aspects of safety to which to attend. In many instances it is found that not enough socket-outlets are provided with the result that operatives using tools powered by electricity have to use very long flexible leads, with cable couplers. Cable couplers themselves, unless maintained properly, can give trouble with faulty connections. Excessive lengths of lead (especially in the flex sizes) often produce considerable voltage drops. In sites investigated some time ago, it was found that because of a shortage of plugs, it was not unusual to find flexible leads to hand tools connected to socket-outlets by wooden wedges driven into the socket contacts.

The main feature of any temporary installation is the provision of electrical services quickly and cheaply. This has too often meant TRS and PVC cables being strung on any convenient support to cut down the cost of the time spent on the job. But the hazards in a temporary installation are far greater than those found in the more permanent installation, and so more care must be taken to see that the installation methods used are those which will go a long way to reduce, as far as is practicable, the dangers arising from shock. Building sites, in particular, offer excellent conditions for shock: wet, damp and exposure of cables to mechanical damage. Thus, great care must be taken with the earthing arrangements, and the positions of lighting fittings, switches and so on. If trailing leads are to be used, then watertight glands must be supplied. And fittings and accessories exposed to the weather, should be of the weatherproof and/or waterproof type.

The result is, that unless careful attention is paid to all aspects of temporary installations, the electrician must spend much time in repairing and maintaining the system, with loss of building operatives' time.

On sites where the electrician is not always available, the use of circuit-breakers is recommended, rather than fuses of the rewirable type. Socket-outlets should be provided with hoseproof, spring-back covers, so that the socket contacts are protected when not in use. All plugs and sockets should have an indication of the working voltage and current, and coloured yellow for clear indication.

In many instances, the cables, wiring accessories and switchgear are often dismantled from one temporary installation to be used on another site. Secondhand equipment must always be tested to ensure its condition; and insulation must always come up to the requirements of the IEE Regulations.

Full requirements for the safe distribution of electricity on construction and building sites is contained in British Standard Code of Practice, CP 1017, to which reference should be made.

21

Circuit-control devices

Definitions and terminology

Switch. A mechanical device for making and breaking, non-automatically, a circuit carrying a current not greatly in excess of the rated normal current of the device.

Contactor. A mechanical device for frequently making and breaking a circuit. It can be operated electro-magnetically, electro-pneumatically, or mechanically as, for instance, a camshaft.

Circuit-breaker. A mechanical device for making and breaking automatically a circuit both under normal and abnormal conditions, such as a short-circuit.

Quick-break switch. A switch in which a quick break is ensured by means of a spring, or otherwise independently of the speed of action of the operator.

Slow-break switch. A switch in which the speed of breaking is dependent upon the speed of action of the operator.

Linked switches. Switches linked together mechanically so as to operate simultaneously or in a definite sequence.

Changeover switch. A switch for changing over from one set of electrical connections to another set of electrical connections.

Reversing switch. A switch for reversing the connections of an electrical circuit.

Time switch. A switch embodying a clock or equivalent mechanism which may be set to make and/or break a circuit at a predetermined time or times.

Close, to. To close a switch or the like. To operate a switch in such a manner as to bring its moveable parts into a position which allows an electric current to flow.

Open, to. To open a switch or the like. To operate a switch in such a manner as to bring its moveable parts into a position which does not allow the passage of an electric current.

Fuse-switch. A switch, the moving part of which carries one or more fuses.

Switchfuse. A unit comprising a switch and one or more fuses. The fuses are not carried on the moving part of the switch.

Relay. A device by means of which an electrical circuit is indirectly controlled by a change in the same or another electrical circuit.

Pole. Applied to a switch or circuit-breaker or similar apparatus to denote the number of different conducting paths or phases which the device opens or closes simultaneously.

One-way. A single path for current.

Two-way. Two alternative paths for current.

Multi-way. Two or more alternative paths for current.

Travel. The distance through which the moving contacts pass between the fully-closed and the fully-open positions.

Tumbler switch. A switch operated by a lever handle on the face of the switch and capable of being rocked in a plane perpendicular thereto. It is usually of small dimensions and suitable for dealing with small amounts of power only.

Surface switch. A switch intended for surface mounting.

Flush switch. A switch intended for mounting behind a switchplate suitable for mounting flush with the surface of a wall.

Semi-recessed switch. A switch with its base suitably formed for recessing partially into the block or box to which it is offered.

Two-way switch. A single-pole changeover switch without an OFF position, generally used where it is desired to control a circuit from two or more positions.

Intermediate switch. A switch for controlling a circuit where more than two positions of control are required, and so called because it occupies an intermediate position between the two two-way switches used in conjunction with it.

Series-parallel switch. A two-position switch giving series connections in one position and parallel connections in the other position.

Ceiling switch. A switch intended to be mounted on the ceiling of a room and operated by a chord.

Bell-push. A push-button switch designed to close an electric bell circuit when the button is depressed, and to open the circuit when the pressure is removed from the button.

Blade. The moving part of a switch which makes contact with the contact-jaw in closing the circuit.

Contact-jaw. The fixed part with which a blade makes contact in closing the circuit.

Dolly. The operating member of a tumbler switch in the form of a lever pivoted on the face of the switch and rocking in a plane perpendicular thereto.

Circuit conditions

Every electrical circuit has its own characteristics, which means that it will show some peculiar electrical property depending on the type of load connected to it. For instance, a circuit which has a purely resistive load (a resistor used as a lamp filament, or heater element) will show a current which rises when the circuit is first switched on and then falls as the element reaches its normal operating condition. This means that the switch or other circuit-control device must at least be able to break the full-load current taken by the resistor. This applies particularly if the circuit has a dc supply. If, however, the supply is ac, when the switch contacts separate there may be a small arc drawn out between the contacts. This characteristic is even more noticeable when the resistor is in the form of a coil (e.g. in a firebar element). This effect is caused by the electrical property which a coil has in an ac circuit. It is called the 'inductive effect' and is explained more fully in Chapter 25.

If, instead of a resistive conductor wound in the form of a coil, a low-resistance conductor is wound round a soft-iron core, the item is then known as a 'choke' or inductor, and the circuit is said to have 'inductive characteristics', which lead to switching problems. A fluorescent circuit is an inductive circuit, as is a motor circuit.

If the circuit has a capacitor included in it, it will also show certain characteristics which may be shown as arcing between switch contacts as they separate. The most pronounced effects of the inclusion of an inductor or a capacitor in a circuit is seen when an ac supply is used. However, small capacitors are often used connected across switch contacts to absorb the sparking caused by contact separation. Used in this way they are sometimes called 'radio-interference suppressors' (e.g. in fluorescent lamp switch starters).

Thus, before a circuit-control device is chosen the circuit to be controlled must be studied so that the device can handle, without damage to itself or the associated circuit wiring, the conditions in the circuit when it is connected or disconnected from its supply. The sections in this chapter which follow indicate the type of control for a circuit which various devices offer.

Contacts

There is in existence an extremely wide range of electrical-contact types used to control the flow of an electric current in a circuit. The action of any pair or pairs of contacts is (*a*) to 'make', to allow the current to flow, and (*b*) to 'break', to prevent the current flow. When this action is contained in a specially-designed wiring accessory or apparatus it becomes one of the many forms of devices used to control circuits: switches, contactors, circuit-breakers and the like.

The basic requirements of any pair of contacts are (*a*) low resistance of the contact material and (*b*) low resistance between the two contact surfaces when they meet to make the circuit.

When these requirements are satisfied, the two main factors which lead to switch troubles are very much reduced. Though one can choose a low-resistance contact material (e.g. copper), one cannot always control the amount of pressure required to keep the two contact surfaces closed sufficiently to reduce what is called 'contact resistance'. A switch, for instance, which is operated many times will eventually reach a stage when its springs become weakened, with the result that pressure of the contacts is lost to such an extent that heat is generated and a breakdown of the switch follows.

The higher the resistance of contact material the more heat (I^2R watts) there will be when a current passes along it. The second factor involved in the design of switch contacts is the amount of pressure needed to keep the two contact surfaces together. All circuit-control devices which meet the relevant specifications of the BSI are tested very rigorously to ensure that they stand up to more wear and tear than they would meet with in normal use. Even so, most contact troubles met with in practice involving the use of circuit-control devices can be traced to insufficient contact pressures.

The material most often used for contacts is copper; this is because it is available in commercial quantities and it has a very low resistance. The terminals associated with the contacts, to which cables and wires are attached, are most often made from brass or phosphor-bronze. These two metals are much harder than copper and so can withstand a certain amount of rough handling with screwdrivers when wiring is being carried out.

The insulating materials used in circuit-control devices include vitrified ceramic (for the bases of switches), bakelite (for switch covers and cases), nylon and mica (for carrying the moving contacts of switches), and insulating oil (used in oil-break circuit-breakers).

In many circuit-control devices silver is used, either as a contact facing, or as the contact itself. The material has a resistance lower than that of copper; it also has high heat-dissipation characteristics and is, for this application, economical to use. Motor-control switches sometimes have contacts of silver-cadmium oxide to reduce the tendency to weld together with heat.

Liquid mercury is also used in special switches called mercury switches. This material has a low contact resistance and a high load-carrying capacity, and can be used in situations with ambient temperatures from about -17°C to 204°C.

Because the contacts are the heart of the circuit-control device, it follows that their surfaces must be kept clean at all times. Cleaning fluids are available for this purpose. Other maintenance points are the periodic tightening up of conductor terminals and connections, and ensuring that springs have not weakened through use, or that cam surfaces have not become worn.

There are two classes of duty for circuit-control devices: (*a*) light current and (*b*) heavy current. Into the first class fall generally lighting switches, relays, and bell-pushes; the second class includes contactors and circuit-breakers.

Switches and switchfuses

N.B. Circuits mentioned in this section are detailed in Chapter 28.

A switch is a device for controlling a circuit or part of a circuit. The control function consists of energising an electrical circuit, or in isolating it from the supply. The type of switch generally indicates the form which this control takes. For instance, a single-pole switch (usually called 'one-way') controls the live pole of a supply. A double-pole switch controls two poles.

A common type of switch in use today is the tumbler type with a rating of 5A, to control lighting circuits. Switches with a 15A rating are also used to control circuits which carry heavier currents on both power (socket-outlet) and lighting arrangements.

Switches are designed for use on dc and/or ac. In a dc circuit, when the switch contacts separate an arc tends to be drawn out between the separating surfaces. This arc is extinguished only when the contacts are far enough apart and when the breaking movement is quick.

Investigation of a dc switch will indicate the length of the gap required when the switch is open. Compare this gap with the gap length on an ac-only switch; it will be found that the latter is very much smaller. The reason for this is that ac tends to be what is called 'self-extinguishing'. In an ac circuit, during the time taken for the contacts to open, the voltage, which is alternating, varies between zero and a maximum. It is at the zero position of the alternating voltage that the arc drawn between the parting contacts of an ac-only switch is extinguished—and it does not establish itself again in normal circuit conditions. Thus, a switch designed for use only on an ac system need have only a small gap and, furthermore, the contact movement does not require to be operated so rapidly as is the case with dc switches.

Quick-make-and-break switches are used for dc circuits. Quick-make, slow-break switches are recommended for ac circuits, particularly where the load is an inductive one, for instance where fluorescent lamps are being used.

The most common lighting circuits are controlled by using single-pole one-way and two-way switches, double-pole switches and intermediate switches. Other types of circuit-control devices and switches are dealt with later in this chapter.

The single-pole, one-way switch provides the ON and OFF control of a circuit—from one position only. When the switch is closed, the lamp is on; when the switch is open, the lamp is off. One-way switches are available with a 'loop-in' terminal mounted on the base of the switch, but isolated from the other parts. This terminal is used for looping the neutral conductor, a wiring method which saves on cable costs by running twin cables only.

The double-pole switch is used in any situation where the voltage of the neutral conductor of a supply system is likely to rise an appreciable amount above earth potential: use of the double-pole switch means that a two-wire circuit can be completely isolated from the supply. The usual application is for the main control of sub-circuits and for the local control of cookers, water-heaters, wall-mounted radiators, and other fixed current-using apparatus. The double-pole switch is often used for the 'master' control of circuits, the switch being operated by a 'secret key' attachment, and in consumer units for the complete isolation of an electrical installation from the supply.

The two-way switch is basically a single-pole changeover switch offering two alternative routes for the passage of the circuit current. These switches are sometimes known as 'landing' switches from the days when their application in the electrical installation was virtually limited to 'one in the hall, and one on the landing upstairs'.

Though the two-way switch is still used extensively for stair lighting, it is also to be found wherever it is necessary to have one or more lights controlled from any one of two positions. They are nowadays to be found in bedrooms (door and bedside), long halls (at each end) and particularly in any room with two entry doors (one at each door).

In design, the switch has four terminals, two of which are permanently connected together inside the switch by a small copper bar on what is called the 'bar' side. One of the bar terminals is blanked off to form a non-separable contact. The switch feed is taken to the other open terminal on the bar side. The two other terminals are connected to the 'strapping wires'. Two-way switches are used in pairs, interconnected so that the switchwire of the light circuit is taken from the open terminal on the bar side of the second switch.

The intermediate switch offers control of a circuit from any one of three positions, the other two positions being at the two two-way switches with which the intermediate switch is most often used. The intermediate wiring circuit is basically a two-way circuit in which the strapping wires are cross-connected by the two ON positions of the intermediate switch. There are three different kinds of intermediate switch, two of which are in common use. It is thus advisable to check the type with an ohmmeter, or bell-and-battery set, because the method of connecting up differs. Figure 21.1 shows the two common forms of connection made within each type of switch.

The application of the intermediate switch in electrical installations has so far been very limited. But there is no reason why it should not be used more extensively. Long halls, corridors and passage-ways with many doors are still wired up for two-way control. For reasonable convenience the light or lights should be controlled from every door and entrance. Thus, the user of this type of circuit can make his way through a house, switching on lights before him, and switching off behind him without having to grope about in the dark.

Two or more intermediate switches can be interconnected into the

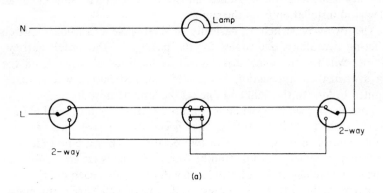

(a)

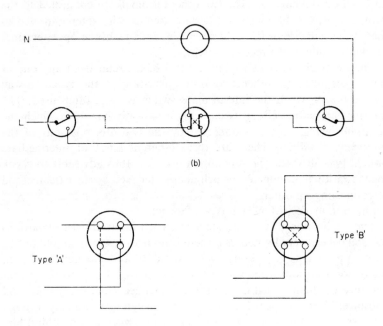

(b)

Type 'A'

Type 'B'

Figure 21.1.

basic two-way circuit to offer control from an almost unlimited number of positions.

The switchfuse is often found as the 'main switch', near the supply-intake position. It is a unit in which the main switch (for installation control) and the main fuses (for the protection of the installation) are combined. In all instances, the switch of the switchfuse cannot be operated when the cover is open, nor can the cover be removed or opened while the switchfuse is closed. The switchfuse, which usually controls a separate distribution board, is of the double- or triple-pole type, depending on the supply system.

Double- and triple-pole switches are found in metal-clad units called isolators. An example is the fireman's emergency switch, painted red and found beside high-voltage gas-discharge lamps such as neons. Isolators are also used to isolate the supply from motors, and heating and non-portable appliances.

Switches are also available in units together with a small fuseboard to form a 'splitter' unit. These are used in small installations for heating and lighting where it is desired to save some expense on main switchgear. The disadvantages are (*a*) it is necessary to switch off all circuits before the lid can be opened to replace a blown fuse; (*b*) the wiring space is very restricted. Thus, the use of the splitter unit should be limited unless expense is the main consideration.

The other instance where a double-pole switch is combined with fuses is the 'consumer unit', which is discussed elsewhere.

The extremely wide range of switchgear types available today can be found in makers' trade literature, study of which is advised so as to become familiar with what is offered for use in electrical installations.

All circuit-control devices, whether switches or other types, must conform to the relevant BS specifications, which thus ensure a minimum guarantee of quality and suitability for use.

Circuit-breakers

The circuit-breaker can be regarded as a switch which can be opened automatically by means of a 'tripping' device. It is, however, more than this.

Whereas a switch is capable of making and breaking a current not greatly in excess of its rated normal current, the circuit-breaker can make and break a circuit, particularly in abnormal conditions such as the occasion of a short-circuit in an installation. It thus disconnects automatically a faulty circuit.

A circuit-breaker is selected for a particular duty, taking into consideration the following: (*a*) the normal current it will have to carry and (*b*) the amount of current which the supply will feed into the circuit fault, which current the circuit-breaker will have to interrupt without damage to itself.

The circuit-breaker generally has a mechanism which, when in the closed position, holds the contacts together. The contacts are separated when the release mechanism of the circuit-breaker is operated by hand or automatically by magnetic means. The circuit-breaker with magnetic 'tripping' (the term used to indicate the opening of the device) employs a solenoid which is an air-cooled coil. In the hollow of the coil is located an iron cylinder attached to a trip mechanism consisting of a series of pivoted links. When the circuit-breaker is closed, the main current passes through the solenoid. When the current rises above a certain value (due to an overload or a fault), the cylinder moves within the solenoid to cause the attached linkage to collapse and, in turn, separate the circuit-breaker contacts.

Circuit-breakers are used in many installations in place of fuses because of a number of definite advantages. First, in the event of an overload or fault, all poles of the circuit are positively disconnected. The devices are also capable of remote control by push-buttons, by under-voltage release coils, or by earth-leakage trip coils. The over-current setting of the circuit-breakers can be adjusted to suit the load conditions of the circuit to be controlled. Time-lag devices can also be introduced so that the time taken for tripping can be delayed because, in some instances, a fault can clear itself, and so avoid the need for a circuit-breaker to disconnect not only the faulty circuit, but other healthy circuits which may be associated with it. The time-lag facility is also useful in motor circuits, to allow the circuit-breaker to stay closed while the motor takes the high initial starting current during the run-up to attain its normal speed. After they have tripped, circuit-breakers can be closed immediately without loss of time. Circuit-breaker contacts separate either in air or in insulating oil.

In certain circumstances, circuit-breakers must be used with 'back-up' protection, which involves the provision of HBC (high breaking capacity) fuses in the main circuit-breaker circuit. In this instance, an extremely heavy overcurrent, such as is caused by a short circuit, is handled by the fuses, to leave the circuit-breaker to deal with the over-currents caused by overloads.

In increasing use for modern electrical installations is the miniature circuit-breaker (MCB). It is used as an alternative to the fuse, and has certain advantages: it can be reset or reclosed easily; it gives a close degree of small overcurrent protection (the tripping factor is 1·1); it will trip on a small sustained overcurrent, but not on a harmless transient overcurrent such as a switching surge. For all applications the MCB tends to give much better overall protection against both fire and shock risks than can be obtained with the use of normal HBC or rewirable fuses. Miniature circuit-breakers are available in distribution-board units for final subcircuit protection.

One main disadvantage of the MCB is the initial cost, although it has the long-term advantage. There is also a tendency for the tripping mechanism to stick or become sluggish in operation after long periods of inaction. It is recommended that the MCB be tripped at frequent intervals to 'ease the springs' and so ensure that it performs its prescribed duty with no damage either to itself or to the circuit it protects.

The circuit-breaker principle is used for protection against earth leakage and is discussed more fully in Chapter 17.

Contactor

When a circuit-breaker has one or more switches in the form of pivoted contact arms which are actuated automatically by an electromagnet, the device is known as a contactor. The coil of the electromagnet is energised by a small current which is just sufficient to hold the pivoted contact arm against the magnet core, and in turn so hold the contacts (fixed and moving) together. Contactors are used in an extremely wide range of applications.

They fall into two general types: (*a*) 'maintained' and (*b*) 'latched-in'. In the first type, the contact arm is maintained in position by the electromagnet. In the latched-in type, the contact arm is retained in the closed position by mechanical means.

Contact design and material depend on the size, rating and application of the contactor. Contactors with double-break contacts usually have silver cadmium-oxide contacts to provide low contact-resistance, improved arc interruption and anti-welding characteristics. Large contactors with single-break contacts use copper contacts for economy. Usually single-break contacts are designed with a wiping action to remove the copper-oxide film which readily forms on the copper tips. Since copper oxide is not a good conductor, it must be eliminated in this way for good continuity.

When the contacts open, an arc is drawn between them. The longer

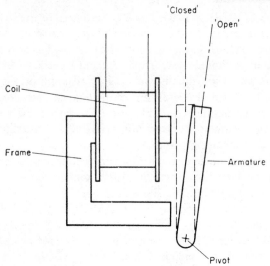

Figure 21.2. Typical contactor arrangement.

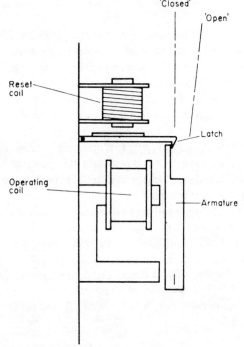

Figure 21.3. Typical latched-in contactor arrangement.

the arc remains, the more the contact material is consumed, and so the shorter is the contact life. The arc can be extinguished by two means: long contact travel, or by use of arc interrupters.

The typical arc interrupter is called a 'blow-out' coil. This uses magnetic means to force the arc and its products away from the surfaces of the contacts, thus lengthening and weakening the arc so that it is eventually extinguished.

Contactors are used to control heating loads, and are often used in conjunction with time-switches and thermostats which close or open the electromagnet current as required. With the contactor, a small current (for the electromagnet) can be used to control a relatively large current in another circuit.

Thermostat

The thermostat is used to control an electric heating appliance or apparatus so that a definite temperature is maintained. It is, therefore, a switch which operates with a change in temperature and is used in the temperature control of rooms, water-heaters, irons, cooker ovens and toasters. It maintains a temperature within defined limits by switching off the appliance when a higher temperature is attained, and switching it on again when a lower temperature has been reached.

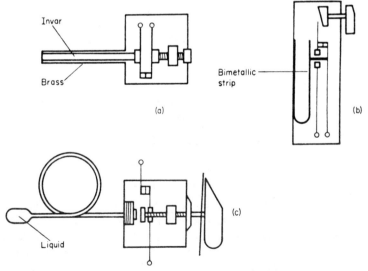

Figure 21.4. (a) Thermostat with metal rod: (b) Thermostat with bimetallic strip: (c) Thermostat with liquid bulb.

The methods used to operate the switch contacts of a thermostat include the expansion of a metal rod, expansion of a liquid or a gas or the bending of a bimetallic strip. Applications of these methods are, respectively, water-heaters, ovens and irons. The illustrations show the basic elements of each type of thermostat.

The speed of response of a thermostat to a change in temperature depends to a large extent on the material used to convey the heat, called the controller. A thermostat whose thermally-sensitive elements are directly opposed to the heat transfer medium will respond faster than one whose elements are shielded by a housing. Liquid-filled systems respond more quickly than gas-filled systems.

Simmerstat

The simmerstat operates by opening and closing a switch at definite time intervals. The ratio of the time the controlled circuit is 'on' to the time it is 'off' is determined by placing a graduated knob at a particular setting. The simmerstat provides a gradual and infinitely variable means of control of, for instance, a cooker hotplate.

The simmerstat consists of a bimetallic strip surrounded by a heating coil. There are two contacts.

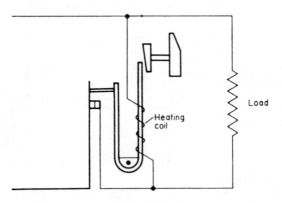

Figure 21.5.

Refer now to Figure 21.5. A control knob operates a cam which varies the setting of the device. When the control knob is in the OFF position, the cam depresses the free end of the bimetallic strip which is in the form of a U, pivoted at the bend. The depression causes the two contacts to open. With the control knob in an intermediate position, the cam allows the free end of the strip to rise and the contacts to close. This completes the circuit of the hotplate and also the heat coil of the simmerstat. After a time, the heat produced by the heat coil causes the bimetallic strip to bend in such a way that the ends of the U move apart. The upper end meets the cam. As it cannot go any further, the lower end continues to take up the movement to travel downwards and so open the contacts. The supply to the hotplate, and to the heat coil, is thus interrupted and will remain so until the strip cools sufficiently to allow the contacts to close again to repeat this cycle.

When the control knob is placed in the ON position, the cam allows the strip to rise freely to the extent that the contacts are never opened. The hotplate now uses its full rated power.

Special switches

With the extensive use of electricity today, it is not surprising to find that there is a great variety of switches and other circuit-control devices with special applications. It is possible to indicate here only some of the most common types.

Three-heat switch

This type of switch is most often associated with the grill-plate of an electric cooker, though it is also used for the heat control of boiling plates. The circuit controlled by the switch consists of two elements of equal resistance. The three-heat switch then offers low, medium and high heat values by its three positions.

Figure 21.6 shows the connections. For low heat, both elements are connected in series to give 25 per cent available power. For medium heat, only one element is connected to give 50 per cent power. For the high-heat condition (full power) the elements are connected in parallel.

The three-heat switch is essentially a rotary or turn switch. The positions are OFF, LOW, MEDIUM, HIGH. The switches are available as a single-pole type (four terminals) or a double-pole type (five terminals).

Time switch

As indicated by its definition, the time switch introduces a 'time element' into an electrical circuit, so that automatic control of the circuit is available at predetermined times. Time switches fall into two general groups: spring-driven and motor-driven. The former uses a mechanism similar to that found in clocks. The latter group uses as the driving unit a small electric (synchronous) motor whose speed is con-

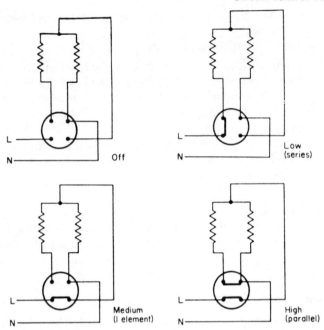

Figure 21.6.

stant and varies only with the 50 Hz frequency of the mains supply. Similar motors are used in electric clocks.

There are many applications for time-switches: shop-window lighting, driveway lighting, street lighting, staircase lighting in multi-tenanted buildings and heating loads, the latter being switched on during 'off-peak' periods when a cheaper tariff is available.

The time-switch control of lighting circuits is often found in such particular applications as poultry houses, where banks of switches control the lighting to simulate summer-daylight conditions and so introduce a 'longer day'. The same technique is also used in horticulture, to hasten the growth of seedlings and plants, particularly during off-season periods of the year.

For normal work, the contacts (either single- or double-pole) are silvered copper, or entirely silver. For heavy currents, mercury-contact time switches are used.

Mercury switch

This is basically a sealed glass tube with a small amount of liquid mercury inside it. Figure 21.7 illustrates a typical switch. The leads are fused into the glass. When the tube is tilted, mercury flows over a second terminal (the first being in permanent contact with the mercury). Thus, contact is made to make the circuit. Mercury switches are made in a very wide variety of types, each type being designed with a particular duty and application in mind.

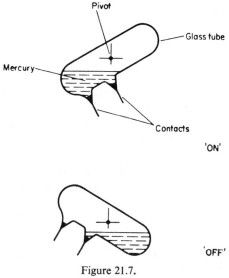

Figure 21.7.

Switches of this type have many advantages: low force required to operate them, low contact-resistance, high load-carrying capacity, low cost, and a long life because of the 'no wear' characteristic of the contacts. It is also relatively insensitive to ambient temperature conditions; a range from −4°C to over 204°C has been specified for some switches. Because the glass is hermetically sealed, the mercury switch is effectively immune to dust, oil and condensation, and can be used where corrosive fumes are present.

Contact connections to the switch are made through flexible leads, or 'pigtails', attached to the embedded electrodes or contacts. Some switches are filled with a reducing gas to keep the surface of the mercury pool free from tarnish.

Because glass is used as the switch container, the contacts are always visible for inspection; and mercury tends to resist heat and arc effects. The materials used for the contacts include tungsten, iron or iron alloys (e.g. nickel-iron) and mercury pools.

Mercury switches are operated by a tilting motion; the method of mounting a switch depends on its application, shape of the actuating member, and the motion produced by it. In the case of a single-throw switch, the glass tube is tilted from the horizontal. Mountings include bimetallic strips, cams and rotating levers. A time-lag element can be introduced by restricting the flow of mercury from one position to another; this is done by a wall placed inside the tube. The wall contains a hole, the diameter of which determines the amount of time-delay.

Rotary switch

The rotary or turn switch offers the facility of controlling a large number of circuits from a local position by using one switch. The three-heat switch is one of the most common examples of the rotary switch. Others include the switches used on switchboards in conjunction with ammeters and voltmeters on three-phase systems to indicate phase-to-phase currents and voltages.

Many banks of contacts can be fitted to a rotary switch so that complete control of circuits is available. Generally the currents are not large: 15A is the usual limit.

Micro-gap switch

This switch derives its type name from the fact that when its contacts (usually silver) are open they are separated by an extremely small gap: anything up to three millimetres. As indicated earlier in the section · on contacts, such switches can be used only on ac circuits. They have many applications apart from 'ac only' lighting circuits.

Thermostats using a 'snap-acting' bimetallic element are in effect micro-gap switches and are to be found in the temperature control of irons, toasters, and cooker heating elements. One industrial application is where a motor overheats and a bimetallic, snap-acting device will switch off the energising current to stop the motor and so protect its winding.

The snap action is always positive in these switches, no matter how rapidly or how gently the force is applied to the operating button. The button can be moved by a plunger, a leaf spring, or a roller and a lever.

Starter switch

Starter switches are used for starting fluorescent lamps. There are two types (*a*) thermal and (*b*) glow, which is now universally used.

The thermal switch consists of two contacts mounted on bimetallic strips. When the lamp is switched off the contacts are normally closed. Beside the contacts is a small heating coil which is connected in series with one of the lamp electrodes (see Chapter 28 for this circuit). When the circuit-control switch is closed, the current flows through the contacts and in the coil of the starter switch. The coil then heats up. The heat is transferred to the bimetallic strip which bends to open the contacts. Because of the inductor (or choke) in the fluorescent lamp circuit, this sudden parting of the contacts causes a high-voltage surge to appear across the ends of the lamp electrodes to start the gas discharge.

The glow-type starter switch consists of two separated bimetallic contact strips contained in a glass bulb filled with helium gas. The contacts are connected to the fluorescent lamp filaments (see Chapter 28). When the circuit-control switch is closed, the mains voltage appears across the two contact strips. This voltage is sufficient to cause a small gas discharge. The heat generated by the discharge effects the bimetallic contact strips which bend forward to meet each other. When they make contact, the current flows through the fluorescent lamp filaments to heat them. The gas-discharge glow in the starter switch now disappears. After a few seconds the bimetallic contact strips cool down and separate. This sudden interruption of the circuit causes a high-voltage surge to appear across the ends of the main lamp electrodes to start the gas discharge.

The voltage which now appears across the contact strips in the starter switch is, during running conditions, insufficient to cause further discharge in the helium gas, and so the contacts remain open while the main lamp is burning.

Two-way-and-off switch

This is a single-pole changeover switch with an OFF position. It is to be found in hotels, ships and hospitals where it is required to have two lamps in circuit while so arranging their control that both cannot be used at the same time.

The two-way-and-off switch can be used as a dimmer control, when in one ON position of the switch only one lamp is lit; in the other ON position, two lamps are connected in series to give a 'dim' light. Other lamp-control arrangements are available when this type of switch is used with other types such as the two-way.

Series-parallel switch

This is a three-position switch with an OFF position when the switch knob or dolly is central. The switch is used to control two points, or two groups of points. In one ON position, the lamp or lamps are connected in series (dim). In the other ON position, the lamp or lamps are connected in parallel (bright). These switches are to be found in hotel corridors, hospitals wards and in railway carriages.

Low-voltage contacts

The most common type of low-voltage contact is the bell-push, which is operated by the direct pressure of a finger on a push-button: the contacts are copper or brass. One is fixed to the base of the bell-push, the other is fixed at one terminal end, its other free end being raised. Pressure on the push-button depresses the contact's free end to complete the circuit. The contacts are usually natural copper, though they are sometimes given a coating of non-oxidisable metal. Other low-voltage contacts use steel springs and phosphor-bronze springs, and are associated with various alarm circuits: burglar, fire, frost, water-level and smoke-density.

Relay

The most common relay is a switch operated by an electromagnet. It consists of an iron-cored coil and a pivoted armature. When the coil is energised, one end of the armature is attracted to the electromagnet and the other end presses two or more contacts together; contacts may also be opened by this movement of the armature.

Relays are either normally-closed (NC) or normally-open (NO). In the first type, when the coil is energised the contacts are open; the contacts close when the coil is de-energised. In the NO relay, the contacts are closed when the coil is energised, and open when it is de-energised. In effect, the relay is an automatic switch.

Relays are normally designed to operate when a very small current flows in the coil. Thus, a small current can be made to switch a larger current on or off, just as a contactor functions from a distant point (remote control). They are also used in bell and telephone systems, and have a wide application in industry.

Other types of relays use a solenoid for their operation. In this instance a plunger is attracted when a predetermined value of current flows in the coil. A time-lag element can be introduced by the addition of an oil- or air-dashpot to delay the movement of the plunger.

Induction and impedance relays operate by the movement of a pivoted disc in the field of an electromagnet; the protective device (usually a circuit-breaker) with which these types are associated is operated by small contacts on the moving disc which, when they close, trip the circuit-breaker. They are used in the protective systems for supply systems, motors, generators and transformers.

The thermal relay consists of a bimetallic strip which heats up when the operating or circuit current flows through it or through an adjacent heating coil. The bending of the strip causes the contacts to either make or break.

Series-circuit controls

Apart from switches and the like which connect or disconnect a supply from a circuit, the circuit can also be controlled, or rather the current in it, by inserting a resistor or inductor in series with the load. The simplest illustration of this is the operation of two 110V lamps in series on a 220V supply. The first lamp in the circuit acts as a 'voltage dropper', dropping 110V so that the balance of 110V appears across the second lamp. The first lamp can quite easily be replaced by a resistor, which will drop or absorb some of the supply voltage to leave a remainder which is sufficient for the operation of the actual load. If the resistor is variable, then the voltage available for the load can be varied according to the load requirements.

The inductor in an ac-series circuit provides much the same facility, and can be made variable by the movement of a soft-iron core into or out of the coil.

Theatre lighting, for instance, requires some form of 'dimmer' control. In the early days this was always done by large banks of variable resistors in series with the lighting load. More modern installations use the variable inductor or 'saturable reactor'.

The main disadvantage in the use of resistors for dimming purposes is that a large amount of heat is generated (I^2R watts) which has to be dissipated. Use of the variable inductor does not present this problem in the same degree.

Series resistors and inductors are also used to limit the amount of current in a circuit. The most common instance of this use is the series choke or inductor in the fluorescent-lamp circuit (or other discharge or arc-lamp circuit). This limiting factor is necessary because, while these lamps are operating, their internal resistance decreases to become so low that a breakdown would occur.

Whether the series device is a resistor or inductor, it must be able to allow sufficient current to pass through it for the normal operation of the actual load. For instance, on a 240V supply, if a load requires 2A for its operation, say a small motor with a rating of 110V, then the series resistor must be able not only to 'drop' the supply voltage to leave 110V for the motor, i.e. to drop 130V, but must also be of sufficient size to pass the 2A current without the resistor heating up unduly.

Protection (physical)

The physical protection of circuit-control devices involves the provision of some method of enclosing the contacts and the switch-movement mechanism. In most instances this is merely providing an insulating enclosure so that the device can be operated with safety from electric shock, hence 'shockproof'. Other enclosures include 'weatherproof' and 'watertight'.

For industrial installations, full protection is often given to the operating dollies of switches by a corner-fixed cover-plate which is both dished and lipped the dolly being located within the dished part of the cover.

Dustproofing is necessary where there is the possibility of fine metal cuttings and dusts, fluff or oil-mists penetrating to the contacts.

Summary of IEE regulations

A8, A9. It is permissible to use a double-pole switch or circuit-breaker to break both neutral and phase simultaneously. But, no single-pole circuit-breaker or switch may be inserted in the neutral pole if the neutral is permanently connected to earth (usually at the supply authority's transformer). As most public supplies in this country have the neutral poles earthed, it will be in exceptional circumstances only that a circuit-breaker or a switch will be permitted in the neutral pole.

All single-pole switches shall be fitted in the same conductor throughout the installation, which shall be the phase or outer of the supply.

A10. This requires that every conductor in an installation shall be protected against excess current by a fuse or circuit-breaker fitted at the origin of the circuit of which the conductor forms part.

A16–20. This requires that the main switchgear associated with three-phase supply systems shall be enclosed in a locked room which is accessible only to authorised persons; or, alternatively that the switch-gear shall be of the interlocked type with the live terminals completely shrouded and a warning-plate on the outside of the case indicating the maximum voltage present.

A23–24. This requires that every final subcircuit shall be derived from a distribution board (or circuit-breaker or switchfuse in the case of one circuit), and that each shall be separate electrically from any other final subcircuit.

D17. Every switch or other electric control shall be placed so as to be out of reach of a person in contact with a bath, basin or pipe-work.

22

Cells

An electric cell is basically two different metallic plates in a conducting liquid (electrolyte). Figure 22.1 shows a simple cell using copper and zinc plates and the direction of current flow in an electrical circuit containing the cell. It should be noted that carbon, though not a metal, is used in Leclanché cells.

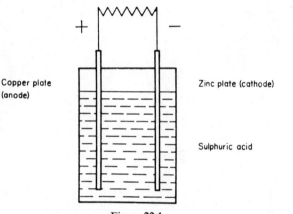

Copper plate (anode)

Zinc plate (cathode)

Sulphuric acid

Figure 22.1.

The current flow is a result of the chemical energy stored within the cell being converted into electrical energy in the external circuit. All cells operate on the same basic principle and the plates may be carbon and zinc, as in the Leclanché cells or two lead plates each having a different active material on their surface as in an accumulator.

Cells are classified in two groups: primary and secondary. The grouping depends upon whether the cell can be recharged electrically or not.

A primary cell is one in which the energy available is obtained from chemical ingredients. When the chemical energy has all been converted into electrical energy the cell is exhausted. This type can be recharged only by renewal of the chemicals.

A secondary cell is 'reversible' in operation. If a current from a direct-current source is passed through the cell the plates undergo a chemical change and chemical energy is stored. This chemical energy can be reconverted as required into electric energy in an external circuit. The process of charging and discharging can be repeated indefinitely.

Primary cells

Leclanché cell

The most common type of primary cell met with in electrical installation is the Leclanché. This cell is available in both wet and dry forms, and Figures 22.2 and 22.3 show the arrangement of each type. The e.m.f. (electromotive force) of both forms is about 1·5V, but the internal resistance of the wet type is about 1 ohm as against 0·3 ohms for the dry type. Because of the high internal resistance and the reduction in voltage due to polarisation the cell is only suitable for supplying low-value intermittent currents. Typical applications are bell circuits and the speech circuits of telephones.

The dry cells are available in the small cylindrical type used in torches, and for bell circuits may be up to about 178 mm high by 76 mm diameter. The voltage is not affected by physical dimensions, but the larger cells supply higher values of current. Wet-type Leclanchés have been used in switchboard protective circuits; the capacity of these was increased by using a zinc plate in cylindrical form surrounding the porous pot.

The wet cell suffers from the disadvantage that it is not portable, but the maintenance is simple and the cell has a long life. During discharge the zinc is attacked by the electrolyte and in the course of time is corroded away. The wet electrolyte tends to creep up the sides of the container and porous pot and crystallise. This is partly prevented by the use of pitch on the upper parts of the container and pot. Any crystals should be removed periodically. The chemical action results in the formation of water in the porous pot and saturation of the carbon and manganese dioxide.

The maintenance of the wet cell is summarised as follows:

(*a*) Renewal of zinc rod and porous pot as required.
(*b*) Maintaining the solution strength by dissolving sal-ammoniac crystals in distilled water.
(*c*) Keeping top and sides of cells clean.
(*d*) Coating terminals with petroleum jelly to prevent corrosion.

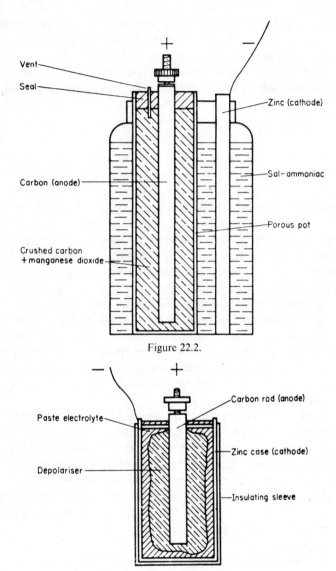

Figure 22.2.

Figure 22.3.

The dry cell is a manufactured article and requires no maintenance. When exhausted the cell tends to swell and the paste electrolyte eats through the zinc casing. This paste attacks any metal nearby and an exhausted cell should be discarded before any damage befalls equipment in which the cell is housed. A leakproof version is available in which a steel jacket covers most of the cell. This is slightly more expensive than the conventional dry cell, but protects equipment against the effects of corrosion.

When voltages greater than 1·5V are required a battery is used consisting of the required number of wet or dry cells connected in series. Small portable equipment such as radios, use dry cells built up in layer-formation to give voltages ranging between 6V and 90V. The capacity of these layer-type batteries is small, being only about a hundred milli-amps (100mA) in the largest sizes.

Mercury cells

In order to supply appreciable currents the Leclanché cell must be made in large sizes. This becomes a problem in some cases due to the weight and the volume occupied by the cell. In cases where fairly high capacity cells are to be housed in a small space use is being increasingly made of mercury cells. These have a current rating of four times that of a dry Leclanché. The cell is available in several different outlines, but an important difference from the Leclanché is that the centre electrode is negative and the outer casing positive. The e.m.f. is about 1·35V.

The mercury cell has many advantages, some of these being outlined as follows:

(*a*) Can be left unattended for very long periods. This makes it suitable for alarm systems and measuring instruments or other equipment which may stand idle for long periods.
(*b*) Does not suffer from polarisation and hence can supply current for a reasonable time.
(*c*) Virtually unaffected by temperature variation.
(*d*) Can withstand momentary overloads.

A possible disadvantage of the mercury cell is its initial cost. Since the life is very much greater than a Leclanché there is little difference, however, in the long-term cost.

In addition to the application given in (*a*) above, the small dimensions of the cell make it most suitable for use in small radios and deaf hearing-aids.

It must be remembered that the mercury cell is a primary cell and cannot be recharged electrically. Maintenance is practically nil, the cell being replaced when exhausted.

Secondary cells

Secondary cells, usually referred to as accumulators, in common use are the lead-acid and alkaline types. These cells have the great advantage that they can be recharged when exhausted from a suitable electricity supply. Their low resistance means they can supply large currents. Provided it is correctly maintained the life of an accumulator in regular use may be several years.

Lead-acid cells

The lead-acid cell consists of either a pair or two sets of lead plates immersed in a solution of dilute sulphuric acid. If current is passed through the cell from a direct-current source, such as a generator or even another battery, chemical changes occur at each plate. The plate connected to the supply positive becomes coated with lead peroxide. The negative becomes a type of spongy lead. Lead sulphate is removed from each plate and dissolves in the acid thus increasing its strength (specific gravity). The conditions for a cell are now fulfilled since we have two different types of lead in an electrolyte. It will be found that if the cell is removed from the supply a voltage will exist between the terminals.

On discharge, that is if the cell is used to supply current to a circuit, the chemical energy stored in the cell is changed back into electrical energy. Both plates tend to turn into lead sulphate and the acid strength drops. In actual practice the accumulator is assumed to be completely discharged when the voltage falls to about 1·85V. If the voltage falls below this the sulphate layer proves difficult to remove by charging and the accumulator is practically ruined.

Modern accumulators are usually of the multi-plate variety. The cell has an odd number of plates sandwiched together but separated from each other by wood or PVC separators. The positives are all joined together and connected to a common terminal. The negatives are connected in a similar manner (see Figure 22.4).

This type of cell has a high current capacity for its size because of the large surface area of the plates. The extra negative plate gives support to the positive plates which would otherwise tend to 'shed' their active material during use. Batteries used in motor vehicles are made up of multi-plate cells connected in series to give the required voltage. These large heavy-duty batteries have individual cells containing up to thirteen plates.

The voltage of a lead-acid cell when charged is about 2·2V falling to

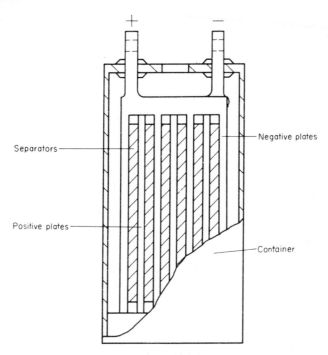

Figure 22.4.

a steady 2V during use. As the cell approaches the end of the discharge period the voltage starts to fall rapidly to 1·85V when it should be recharged. During use the plate colour changes; this change can be used as a guide to condition. When charged, the positive plate is a rich brown; the negative is slate-grey. In addition, bubbles can be seen rising from the plates (gassing). On discharge the plates both tend to become lighter in colour.

Applications of lead-acid cells range from miners' cap-lamps, using small portable cells to large emergency service storage batteries such as used for stand-by lighting purposes. These large batteries may give voltages of up to 110V by connecting a number of cells in series and are of heavy-duty rating.

The life of a lead-acid cell depends on efficient and regular maintenance and the principal points to observe are as follows:

(*a*) The currents on charge and discharge should not exceed those stated by the manufacturers.

(*b*) The cell should be recharged when the voltage approaches 1·85V or when the specific gravity falls to about 1·18. The specific gravity is checked using a hydrometer and drawing a sample of acid from the cell.

(*c*) The level of the electrolyte must always be kept above the plates. If it is required to add liquid, to replace electrolyte lost by evaporation, only distilled water should be used. Acid of the correct strength may be added to replace any lost by spillage.

(*d*) Cells making up stand-by supplies (e.g. emergency lighting, switch-board protection supplies, etc.) should receive a trickle charge from a suitable dc supply.

(*e*) The tops of the cells should be kept clean and free from any liquid. The terminals should be lubricated with petroleum jelly to prevent corrosion.

(*f*) Any cell withdrawn from service and stored should be examined and given a charge at regular intervals. The current value for charging in this case is just a little over half that for normal use.

In stations where batteries are charged, provision should be made for adequate ventilation. Hydrogen gas is liberated during charging and no naked light must be used in any charging stations.

Alkaline cells

There are two types of alkaline cell in use; the nickel-iron and the nickel-cadmium. Both types are housed in steel cases and the electrolyte is caustic potash (potassium hydrate) and distilled water to give a specific gravity of about 1·2. The e.m.f. of these cells is about 1·4V and the electrolyte strength does not change during operation.

In the nickel-cadmium cell the positive material is nickel hydroxide enclosed in pockets made from finely-perforated steel ribbon. The negative material is cadmium oxide enclosed in similar steel-ribbon plates. This gives a flat plate formation. The nickel-iron cell differs in that the active materials are enclosed in steel tubes and iron oxide is used for the negative material.

Alkaline cells have the advantages of long life, saving in weight over the lead-acid cell, are virtually unaffected by momentary overloads and are very strong mechanically. They are, however, more expensive than acid cells although this is partly offset by their long life. Maintenance is similar to that of lead-acid cells, but not quite as critical. For example, an alkaline cell can discharge almost completely without ill effect, although this is naturally bad practice. One point to be borne in mind is that **the electrolyte is very corrosive** and can inflict serious damage if allowed to come into contact with the skin. Alkaline batteries are made up by connecting cells in series. The steel casings of the cells are electrically negative and batteries are constructed by housing each cell in a wooden or plastic insulating case.

Alkaline batteries can be used in practically any situation to replace acid batteries. Typical applications are batteries used in transport, emergency lighting systems and in miniature forms in rechargeable torches and transistors.

The main points to be observed regarding storage batteries are summarised as follows:

(*a*) Reverse-current protection is to be provided where dc generators are connected in parallel with a battery. Effective means are to be provided to protect the batteries from excessive charging or discharging current. This protection can be a fuse or a circuit-breaker.

(*b*) Switchboards controlling the supply from a secondary battery installation must have a double-pole switch or circuit-breaker to isolate the battery from both load and charging circuits. If the charging circuit is used to share the load with the battery, means must be provided to isolate the charging circuit from the battery and the load. This can be a double-pole linked switch or circuit-breaker.

(*c*) Batteries should be arranged to be accessible from at least one side and the top. If the working voltage of the battery is 60V or more, the individual cells must be supported on glass or porcelain insulators. The battery racks must be insulated if the working voltage is above 120V. All connecting bolts should be lubricated with petroleum jelly unless of the non-corrosive type. Open-type cells to be fitted with spray-arresters to minimise the effects of acid spray particles in use.

(*d*) Celluloid shall not be used other than for portable batteries, and in the case of batteries using celluloid construction the charging arrangements shall be such to minimise the effects of fire.

(*e*) Rooms containing secondary batteries shall be adequately ventilated to the outside air. If the batteries use sulphuric acid, the fittings must be non-corrosive or treated with acid-resisting paint.

It should be noted that reverse-current protection is required when charging is from a dc generator. Should the driving motor or engine fail the batteries could send current in the 'reverse' direction through the generator causing it to 'motor' thus reversing the field polarity. Subsequent restoration of the supply would cause the generator to give out a voltage reverse to that previously obtained.

23

Electric lamps

Terms and definitions

Filament lamp. A lamp in which a metal, carbon or other filament is rendered incandescent by the passage of an electric current.

Vacuum lamp. A filament lamp in which the filament operates *in vacuo.*

Gas-filled lamp. A filament lamp in which the filament operates in an inert gas.

Arc lamp. An electric lamp in which the light is emitted by an arc.

Discharge lamp. An electric lamp in which the light is obtained by a discharge of electricity between two electrodes in a gas or vapour.

Lumen. (lm) unit of luminous flux (or 'amount of light') emitted from a source.

Filament lamps

Filament lamps fall into a group of light-producing devices called 'incandescents'. They give light as a result of heating the filament to a very high temperature. Another name for this group is 'temperature radiators', and examples include the torch, the candle, the oil lamp and the gas lamp. It was in 1840 that Groves announced the first incandescent lamp. The first lamp consisted of a platinum spiral operating in a glass tumbler inverted over water; this arrangement was presumably to minimise the effects of draughts on the glowing spiral and to prevent rapid burning of the filament due to the oxygen in the atmosphere. The first British patent for an incandescent lamp was granted in 1841. In 1860, Sir Joseph Swan produced the first lamp using a carbonised paper strip: later, carbonised filaments from silk were used. Until 1900, the carbon-filament lamp enjoyed an undisputed field of use; at this date metal filaments were just coming into use, and by 1910 they had superseded the carbon lamp. The carbon-filament lamp is still used today, though only for special purposes. It gives a reddish light and operates at a temperature of about 2,000°C. Above this limit, the carbon evaporates rapidly and blackens the glass bulb or envelope. The output of light from a carbon filament lamp is about 3 lm/W (lumens per watt).

Tungsten-filament lamps operate at a temperature of about 2,300°C, and have a light output of 8 lm/W. The filaments are made from fine-drawn tungsten wire. The first lamp to use tungsten (about 1910) had the air evacuated from the bulb—the so-called vacuum lamp. Later, in 1913, the bulb was filled with argon and nitrogen which are inert gases and do not support combustion. This development enabled the filament to be operated at a higher temperature without the undue evaporation which tends to take place in a vacuum. The operating temperature of the gas-filled lamp is about 2,700°C and the light output is about 12 lm/W. The early lamps had a single coil filament. Later, the coiled-coil lamp was produced—the coiled filament was itself formed into a coil. The light output of a coiled-coil lamp is about 14 lm/W. The main advantages of the coiled-coil filament lamp are (a) the filament has a more compact formation and (b) the heat losses due to convection currents in the gas are reduced, so giving a higher light-output efficiency.

Incandescent lamps are used for many purposes and are available with many variations. 'Pearl' lamps have the glass bulb internally frosted. Other types have the glass bulb silica-coated internally. The following are some of the applications of filament lamps:

General lighting. Vacuum or gas-filled lamps; top-reflector lamps (frosted and half-mirrored inside).

Festive lighting. Spherical, candle and chandelier lamps, with frosted or opalised finish; coloured and white.

Spot and flood lighting. Lamps made from pressed glass and internally mirrored to radiate a definite beam of light. The floodlight has a relatively broad beam. The spotlight has a narrow beam. These lamps are very strong and are used for shop and showcase illumination. Floods are used for outdoor illumination (gardens, parks and sports grounds).

Reinforced construction lamps. These lamps are intended for use where vibration and shock are more than normal: usually called rough-service lamps.

Signal lamps. These are used for signalling purposes on switchboards and for indication installations: sometimes called 'pygmy' lamps.

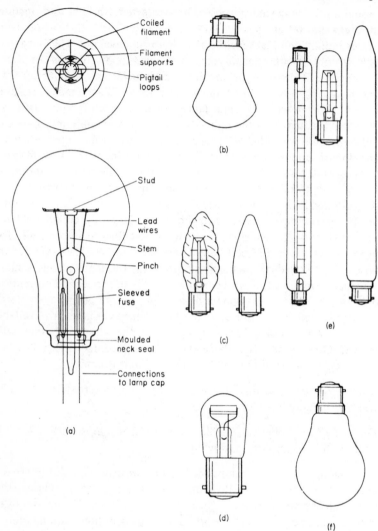

Figure 23.1. (a) The filament lamp and its construction: (b) The 'mushroom' filament lamp: (c) Candle filament lamps for decorative lighting: (d) Neon lamp: (e) Various types of tubular filament lamps: (f) Modern filament lamp.

Thermal radiation lamps. These lamps are used in piglet- and chicken-rearing. Heating lamps are hard-glass bulbs internally mirrored for use for short periods at a time. They are used for heating in bathrooms and in industry for drying processes (e.g. stove-enamelling).

Special lamps. In this group are included sewing-machine lamps, oven lamps, showcase lamps and Christmas-tree illumination sets.

One type of lamp of recent introduction is the quartz-iodine lamp, in which iodine vapour is used to control the rate of evaporation of the filament material and thus prolonging its life. This type of lamp is smaller than other types of filament lamps, though the problem of heat dissipation is increased. Usually the metal housing of the fitting is of finned construction, and the terminal chamber for cable entry is partially separated from the main housing. The main application of this lamp is for floodlighting. The reflectors must be protected by toughened glass, because there can be a considerable difference in temperature between the edge and the centre.

Discharge lamps

Under normal circumstances, an electric current cannot flow through a gas. However, if electrodes are fused into the ends of a glass tube, and the tube is slowly pumped free of air, current does pass through at a certain low pressure. A faint red luminous column can be seen in the tube, proceeding from the positive electrode; at the negative electrode a weak glow is also just visible. Very little visible radiation is obtainable. But when the tube is filled with certain gases, definite luminous effects can be obtained. One important aspect of the gas discharge is the 'negative resistance characteristic'. This means that when the temperature of the material (in this case the gas) rises, its resistance decreases—which is the opposite to what occurs with an 'ohmic' resistance material such as copper. When a current passes through the gas, the temperature increases and its resistance decreases. This decrease in resistance causes a rise in the current strength which, if not limited or controlled in some way, will eventually cause a short circuit to take place. Thus, for all gas-discharge lamps there is always a resistor, choke coil (or inductor) or leak transformer for limiting the circuit current. Though the gas-discharge lamp was known in the early days of electrical engineering, it was not until the 1930s that this type of lamp came onto the market in commercial quantities. There are two main types of electric discharge lamp:

(*a*) Cold cathode.
(*b*) Hot cathode.

The cold-cathode lamp uses a high voltage (about 3·5kV) for its operation. For general lighting purposes they are familiar as fluorescent tubes about 25 mm in diameter, either straight, curved or bent to take a certain form. The power consumption is generally about 8 watts per 30 cm; the current taken is in milliamps. The electrodes of these lamps are not preheated. A more familiar type of cold-cathode lamp is the neon lamp used for sign and display lighting. Here the gas is neon which gives a reddish light when the electric discharge takes place in the tubes. Neon lamps are also available in very small sizes in the form of 'pygmy' lamps and as indicating lights on wiring accessories (switches and socket-outlets). This type of lamp operates on mains voltage. Neon signs operate on the high voltage produced by transformers.

The hot-cathode lamp is more common. In it, the electrodes are heated and it operates generally on a low or medium voltage. Some types of lamp have an auxiliary electrode for starting. Other types, more within the scope of this chapter, called 'switchstart' lamps, use a switching arrangement in the circuit (see Chapter 21).

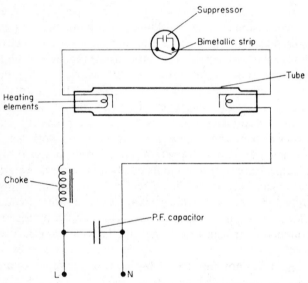

Figure 23.2. Fluorescent-tube starting circuit.

The most familiar type of discharge lamp is the fluorescent lamp. It consists of a glass tube filled with mercury vapour at a low pressure. The electrodes are located at the ends of the tube. When the lamp is switched on, an arc-discharge excites a barely-visible radiation, the greater part of which consists of ultra-violet radiation. The interior wall of the tube is coated with a fluorescent powder which transforms the ultra-violet rays into visible radiation or light. The type of light (that is, the colour range) is determined by the composition of the fluorescent powder. To assist starting, the mercury vapour is mixed with a small quantity of argon gas. The light produced by the fluorescent lamp varies from 35 to 45 lm/W. The colours available from the fluorescent lamp include a near-daylight and a colour-corrected light for use where colours (of wool, paints, etc.) must be seen correctly. The practical application of this type of lamp includes the lighting of shops, domestic premises, factories, streets, ships, transport (buses), tunnels and coalmines.

The auxiliary equipment associated with the fluorescent circuit includes:

(*a*) the choke, which supplies a high initial voltage on starting (caused by the interruption of the inductive circuit), and also limits the current in the lamp when the lamp is operating.

(*b*) The starter; types are described in Chapter 21.

(*c*) The capacitor, which is fitted to correct or improve the power factor by neutralising the inductive effect of the choke.

The so-called 'switchless' start fluorescent lamp does not require to be preheated. The lamp lights almost at once when the circuit switch is closed. An auto-transformer is used instead of a starting switch.

The IEE Regulations relating specifically to discharge lamp circuits are detailed in Section G. The main requirements, so far as fluorescent lamps are concerned, include the preferred use of quick-make-slow-break circuit switches, the rating of circuit switches (G2) and the rating of the final subcircuits supplying the lamps (G3).

Lamp caps

There are many means used to connect lamps effectively to the supply. The most common are for General Lighting Service (GLS) and fluorescent lamps.

Bayonet cap (BC). For filament lamps up to 150W.

Edison screw (ES). For filament lamps 150W to 200W.

Goliath Edison screw (GES). For filament lamps from 300W to 1,500W.

For fluorescent lamps there are the Bi-pin and the Bayonet cap.

Lamp circuits and controls

For the various type of lamp circuits, and their control and ancillary equipment see Chapter 28 of this book.

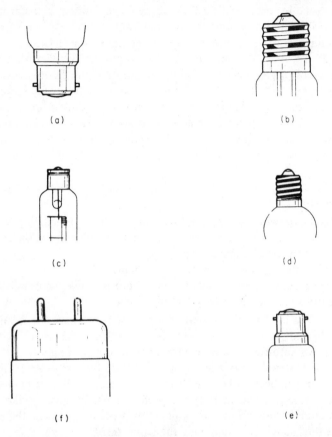

Figure 23.3. Types of lamp caps: (a) Spigot-type bayonet-cap: (b) Large Edison Screw: (c) Non-spigot centre-contact: (d) Small Edison Screw, centre contact: (e) Small bayonet-cap, spigot type: (f) Bi-pin type (fluorescent tube).

Lamp ratings

All lamps are identified by their 'wattage', or the quantity of electrical energy they take to produce a definite quantity of light. Note, however, that carbon lamps are rated in 'candle-power'. This term is a flashback to the days when the standard light-producing device was a candle made from spermicetti wax. Nevertheless, filament lamps are rated at so many watts at a specified or rated voltage. The light output of lamps is very dependent on the supply voltage. Taking a light output of 14 lm/W, a 60W, 240V lamp will produce 14×60 lm $= 840$ lm. The current taken is $I = W/V = \dfrac{60}{240} = 0{\cdot}25A$. At this current we can

calculate the resistance of the filament, $R_f = \dfrac{V}{I} = \dfrac{240}{0{\cdot}25} = 960$ ohms.

Suppose now the supply voltage was increased to 260V. The current would rise to $I = \dfrac{V}{R_f} = \dfrac{260}{960} = 0{\cdot}27A$, and the wattage taken to $W = VI = 260 \times 0{\cdot}27 = 70W$. The light output would also increase to $14 \times 70 = 980$ lm. The opposite would, of course, occur if the supply voltage were to fall below the rated voltage. Briefly, then, the light output depends on the voltage of the supply. All filament lamps have a nominal life of 1,000 hours, that is, each lamp will burn at rated voltage for an average of 1,000 hours before failing. To 'overvolt' a lamp (by increasing the supply voltage or by connecting, say, a 240V lamp to a 260V supply), will increase its light output, but reduce its useful life. Even overvolting by 5 per cent (e.g. a 230V lamp on 242V) will halve its life. Undervolting a lamp will reduce its light output and prolong its life; but this does not result in a corresponding reduction in the wattage consumed and electricity is in effect being wasted.

The nominal life of a fluorescent lamp is between 5,000 hours and 7,500 hours. With switched or switchless starting gear, the control gear losses generally amount to about 15 per cent of lamp watts. Thus, the circuit of an 80W fluorescent lamp will take from the supply a total of 92W.

Light measurement

The amount of light falling on a surface is measured (in lumens per m^2; Unit = lux) by an instrument called a photometer or light-meter. It consists of a cell made from three layers of metal: (*a*) a transparent film of gold; (*b*) a film of selenium; (*c*) a steel baseplate.

A connecting ring makes contact with the transparent film. Another connection is taken from the steel plate. These connections are taken to a very sensitive moving-coil instrument, which has a scale graduated in $1m/m^2$. When rays of light fall on the surface of the cell between the gold and selenium films, electrons are freed, to cause a current to flow in the moving-coil of the meter movement. This current is approximately proportional to the amount of light falling on the cell. The instrument is used to check that the amount of light falling on working planes (tables, desks, benches) is sufficient for a particular job to be

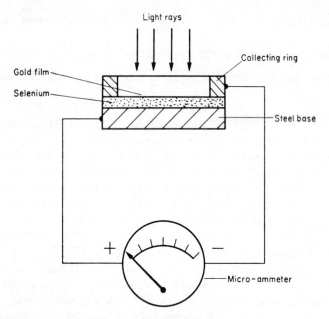

Figure 23.4. The light meter.

done with no strain on the eys of the worker. The Illuminating Engineering Society publishes Tables indicating the optimum amount of light
required to perform various tasks in industry and in the home.

24

Electromagnetic devices

In electrical engineering work, extensive use is made of various types of electromagnetic devices for providing a mechanical force to operate different kinds of mechanism. Operation by electromagnetic means gives the advantage of automatic control with possible remote control. There are three main types of electromagnet:

1. *Tractive type*. In this type, the electromagnet attracts an armature to which the load mechanism is attached. An example is the contactor or relay, where the attraction and release of the armature activates one or more sets of contacts.

2. *Solenoid type*. In this type, the operating coil surrounds a sliding core or plunger which is drawn into the coil, usually in a vertical direction, when the coil is energised.

3. *Lifting type*. In this type, the poles of the electromagnet are brought into contact with magnetic material so that the material can be transported.

In addition to these types, there are the applications in which the electromagnet forms an integral part of a complete mechanism, such as a magnetic brake or clutch. The underlying principle is the same in all cases—based on the laws of electromagnetism applying to the attraction and repulsion of magnetised surfaces.

Solenoids

The simplest form of electromagnetic solenoid is a cylindrical coil, within which the bore of an iron plunger is free to move. The pull extended on the plunger with this simple arrangement when the coil is energised is very small. The pull is increased by making the coil long and increasing the cross-sectional area of the plunger. Another method of increasing the pull of the solenoid is to provide the solenoid with an iron circuit to complete the magnetic path as far as possible with the

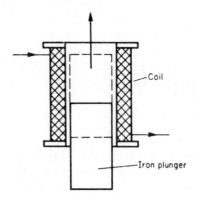

Figure 24.1. Solenoid and plunger.

exception of a working air-gap. This iron circuit is generally a frame bent to shape from steel strip. In this form, the solenoid has a wide application for overcurrent and low-voltage relays. In an ac solenoid, the flux provided by the coil passes through a maximum and falls to zero twice in each cycle, so that the pull is not constant. Thus, an ac solenoid is not so efficient as a dc type.

Electromagnets

The general form of the electromagnet is shown in Figures 24.2. and 24.3. There is a cylindrical pole with shouldered ends which serve to keep the coil in position, a rectangular yoke to which the pole is screwed or bolted and a rectangular armature. Two exciting coils are used; they are connected to give opposite polarities at the respective pole-ends.

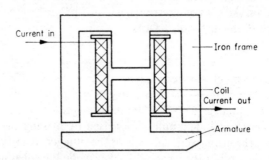

Figure 24.2. Electromagnet, single coil.

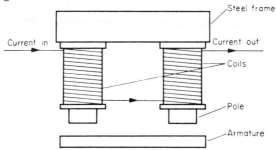

Figure 24.3. Electromagnet, double coil.

Electric bells

There are four types of electric bell in common use:

1. Single-stroke bell
2. Trembler bell
3. Continuous-ringing bell
4. Polarised bell.

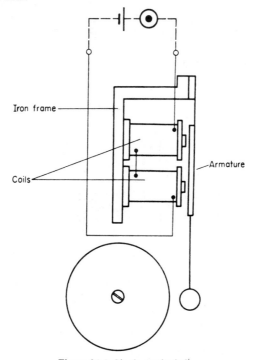

Figure 24.4. Single-stroke bell

1. *Single-stroke bell.* See Figure 24.4. This type of bell, as its name implies, will sound a single note and is used for signalling purposes (e.g. in buses). The two coils of the electromagnet are formed from varnish- or cotton-insulated copper wire and mounted on an insulated bobbin. The bobbins are fitted onto a V-shaped iron core. A soft-iron strip is riveted to a spring-loaded armature, which carries the striker. The gong is hit once only by the striker each time the bell-push is pressed. When the push is operated, the current flows through the coils. The armature is attracted and its striker hits the gong. For correct operation, the push must be pushed ON and released OFF in a quick movement. Otherwise, if the circuit is energised for too long a period, the sound of the note may be muffled because the striker will be held against the gong. As soon as the push is released, there is a break in the circuit and the armature and striker fall back to the original position under the influence of the spring. The double-note of the door-chimer is a variation of the single-stroke bell. Tubular chimes replace the gong. The second note or chime is caused by the rebound of the striker.

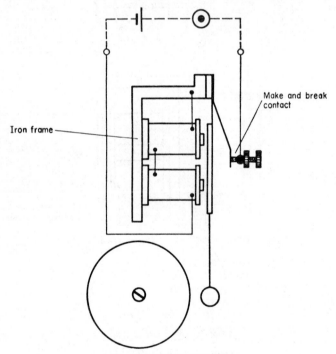

Figure 24.5. Trembler bell

2. *Trembler bell.* See Figure 24.5. This type of bell is in very common use as a door bell. It is similar in construction to the single-stroke bell except that a make-and-break arrangement is provided in the circuit. Instead of the coils being connected directly across the supply through the bell-push, the coils are in series with an adjustable contact-screw against which an armature spring-leaf contact normally rests. When the push is pressed, the current flows through the adjustable contact-screw, the armature and the coils. The coils become energised and attract the armature. This movement of the armature away from the contact-screw breaks the circuit. The coils are de-energised and the armature falls back to its original position. Again contact is made with the screw and the circuit is completed once more. This sequence of movement occurs rapidly. The striker at the end of the armature hits against the gong to cause a continuous ringing sound. The action of the bell is thus a continual making and breaking of the circuit to make the armature 'tremble' between the coil cores and the contact-screw. The ringing will continue for as long as the push is pressed. Because sparking occurs at the contact

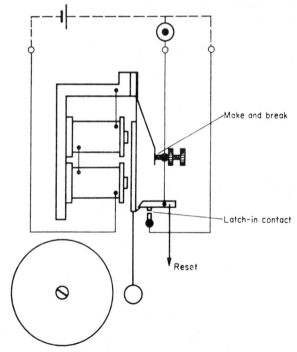

Figure 24.6. Continuous-ringing bell

points, these are often made of silver or platinum, which do not oxidise easily. The tone of the sound produced by the bell can be changed by altering the setting of the adjustable contact-screw.

3. *Continuous-ringing bell.* Figure 24.6. This type is basically a trembler bell, but with either a mechanical or electrical arrangement to make the bell continue to ring after the bell-push has been released. A small lever is placed below the contact-screw. On the first movement or stroke of the armature, this lever drops automatically and, as it drops, it short-circuits the bell-push, thus causing the bell to ring continuously. The lever can be reset by pulling a cord so stopping the bell.

4. *Polarised bell.* Figure 24.7. This type of bell operates from an alternating current obtained from the type of magneto as used in certain types of telephone circuits. It is sometimes called a magneto bell. It consists of a three-limbed permanent magnet with its two outside limbs surrounded by coils wound in opposite senses. A soft-iron armature carrying a striker knob (between the two gongs) is pivoted in the centre, within the magnet field. When the current in the coils has one polarity (e.g. a positive direction), one pole of the magnet is strengthened while the other is weakened. This causes the armature to pivot so that the striker hits one gong. When the current direction reverses, the opposite

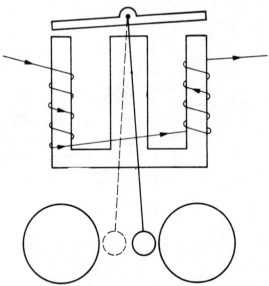

Figure 24.7. Polarised bell

occurs so that the other gong is struck. As each cycle of alternating current is produced, the armature pivots, and the bell produces a ringing sound.

Indicators

Often a bell circuit is arranged so that the bell can be operated from any one of two or more positions. In such circuits it becomes necessary to instal an indicator board to show which bell-push has been operated. There are three main types of indicator:

1. *Pendulum type.* The movement of this type is similar to that of the single-stroke bell. A soft-iron armature is pivoted or hinged at one end; the other end carries a flag. The armature is located in front of the electromagnet, the coil of which is connected in series with a bell-push. When the push is pressed, the electromagnet becomes energised and it attracts the armature. As the adjustable contact-screw on the associated bell makes and breaks the circuit, so does the electromagnet become energised and de-energised. And the armature swings to and fro in pendulum fashion for a short time before coming to rest. The movement of the flag is seen through a clear circular part of the glass screen. The disadvantage of this type of indicator is that if the person called happens to be out of the room when the bell rings, the pendulum may well have stopped swinging by the time the person returns and reaches the indicator.

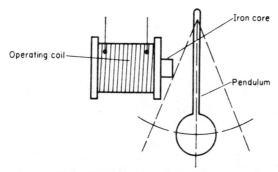

Figure 24.8. Pendulum indicator unit.

2. *Mechanical-replacement type.* In order that some indication that a bell has rung may be displayed for an indefinite time on the indicator board, it is necessary to have some arrangement for the manual replacement of the indicator flag after the request for attention has been satisfied. One method used is by mechanical means.

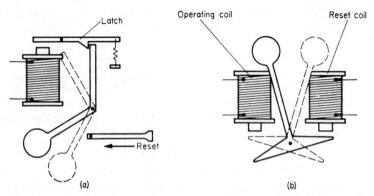

Figure 24.9. (a) Flag indicator unit, mechanical reset: (b) Flag indicator unit, electrical reset.

The mechanical-replacement type of indicator element has a single-core electromagnet. The armature is pivoted near its centre and held away from the magnet core by a spring at the end remote from the magnet. The flag is attached to one end of a pivoted arm which has an extended piece caught normally by a catch on the armature. When the bell-push is pressed, the electromagnet is energised and attracts the armature. The movement of the armature causes the catch on the pivoted arm to be released and the flag moves to show itself in the appropriate space in the window of the indicator board. The flag will remain in this position even though the bell-push is released. To reset the flag, a lever is either pushed or turned round by hand.

3. *Electrical-replacement type.* This type of movement has two separate electromagnets. One coil is in series with the appropriate call bell-push as before. The other coil is connected to a 'replacement' circuit with a separate bell-push. The armature, to which the flag arm is directly attached, is pivoted at its centre in such a way that it can be attracted by either magnet. When the call push is pressed, the armature is attracted to the alarm or indication position. The flag shows in the window of the indicator board. When the replacement push is operated, the armature is attracted by the other magnet and the movement causes the flag to move out of sight, to its original position.

Bell Relays
When a bell circuit is long in length, it may be found that the volt drop along the circuit wires is so great that the bell will not operate, or else does so inefficiently. Additional cells or batteries may be used to raise

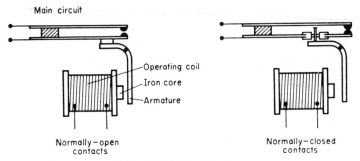

Figure 24.10. Typical relays.

the circuit voltage, to compensate for the lost volts. But this may not always be a satisfactory solution to the problem. The other solution is to employ a relay. The relay is basically a pair of electrical contacts operated by an electromagnet. A very small current is sufficient to energise the electromagnet and attract a spring-controlled armature. The movement of the armature causes the two contacts to close and operate the bell. The relay circuit may be energised separately from its own supply, or it may use the bell supply. Operation of the bell-push will energise the relay and cause the bell to ring. Typical relay circuits are shown in Chapter 28 of this book.

Bell transformers

The simplest bell circuit uses a cell or battery for its source of energy. The bell may also be supplied from a double-wound transformer. It consists of a soft-iron core with two windings: the primary and the

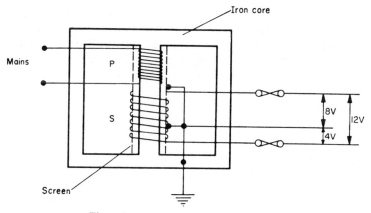

Figure 24.11. Bell Transformer, Class A.

secondary. The primary winding accepts the mains ac voltage. The secondary winding, which has no electrical connection with the primary, produces (by mutual induction) a low voltage. Class A bell transformers provide a choice of three secondary voltages: 4V, 8V or 12V. Class B transformers provide a single secondary voltage of 6V.

Regulations for bell circuits

The following points are to be observed when installing bell circuits operating on extra-low voltage supplies. Mains-operated bells are installed in accordance with the normal mains-supply circuit wiring requirements.

1. The bell transformer must be double-wound.
2. One point of the secondary winding, the core of the transformer and the metal case (if used) must be earthed.
3. Preferably, the transformer should be connected to a separate way in a distribution board. Otherwise, it should have a separate control switch.
4. The cables used to supply the transformer should be of a grade suitable for the supply voltage in use.
5. The secondary wiring need be insulated for extra-low voltage only, provided that it can be completely segregated from mains-voltage power and lighting cables. If the bell-circuit wiring must be run in the same conduit, trunking or duct as power or lighting cables, then the bell-circuit wiring must be of a grade suitable for the highest voltage present in the power and lighting cables.

Because the voltage of bell circuits is very small, the danger of fire is also small. Thus, lightly-insulated wires are used for bell-circuit wiring: generally tinned-copper wire of $1\cdot0\,\text{mm}^2$, $1\cdot5\,\text{mm}^2$ or $2\cdot5\,\text{mm}^2$. Insulation consists of a thin layer of rubber with a double covering of cotton or silk. Where protection is needed, a lead-alloy sheath is provided. Dry joints are not satisfactory. All joints should be properly made and soldered and insulated. Wires are available as single-core or twin-core (usually parallel-twin). The wiring may be run on the surface of walls without further protection. Fixing is usually by means of fibre-insulated staples.

25

Inductors and transformers

An inductor is a coil, usually wound on an iron core, which when connected in an alternating current circuit offers opposition to current flow. The types of inductor met with in installation work are called 'chokes'. A transformer is a device for changing the values of voltage in an alternating current circuit. Both pieces of apparatus depend for their operation on induced voltages set up by a fluctuating magnetic field due to the ac supply system. A brief indication of the principles involved is given below.

Self-induction

When a conductor, or coil, is moved in a magnetic field, an e.m.f. is set up in the conductor. The e.m.f. is only present as long as the conductor is moving. Similarly, the conductor can be stationary and the magnetic field altered, giving the same effect. On any circuit carrying current, a magnetic field is set up due to the current. Figure 25.1 shows a coil carrying current and the resulting magnetic field.

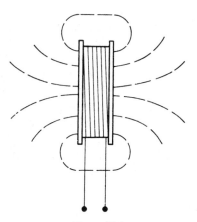

Figure 25.1.

The magnetic field alternates with the same frequency as the current and embraces the coil-conductors. The change in field causes an alternating voltage to be induced in the coil. This induced voltage acts in a direction opposite to that of the supply voltage, and as a result reduces the effective voltage of the circuits so limiting the current. The coil therefore has a 'choking' effect, and offers an opposition to current flow called the Inductive Reactance.

Any circuit in which a change in current produces an induced e.m.f. possesses 'inductance' and the circuit is an 'inductive' one. The voltage induced in this case is due to changes in a single circuit and is referred to as being 'self-induced'.

When an inductive circuit is used on direct-current and is switched on or off, the sudden change in field strength causes a self-induced e.m.f. to be generated. The induced e.m.f. can be greater than the applied voltage depending upon how great is the rate of change of field. This induced voltage can show itself as arcing at the contacts of switches. Direct-current tumbler switches have a faster action and wider contact clearance than ac switches of equal rating to minimise effects of arcing.

Applications

The problem of arcing is not so critical on ac, but voltage surges are still produced on switching. This effect is put to use in the starting of fluorescent discharge lamp circuits. A choke is connected in series with the lamp and operation of the starting switch causes a momentary high voltage to be induced. The high voltage initiates the discharge and allows the lamp current to flow. Once the lamp strikes, the choke then limits the circuit current to the value required to operate the lamp. Discharge lamp circuits and a description of these lamps is given in Chapter 23.

Series chokes, or inductors, are also used in dimmer circuits to reduce the intensity of illumination in stage lighting, etc. The resistance of a choke is small compared to its inductive reactance when used on ac. The reduction in voltage across the lamps is achieved without the power loss and heating which occurs in a resistor dimming circuit. It must be kept in mind, however, that the power factor of the circuit is reduced by the use of the highly inductive choke.

The requirements of the Regulations applying to the installation and operation of chokes are given at the end of this chapter.

Mutual Induction

Consider two coils, wound on a common iron core and one of them connected to the supply. If the switch is closed a magnetic field links both coils as shown in Figure 25.2. The opening or closing of the switch will cause a self-induced e.m.f. to be generated in the coil connected to the supply. Since the field also links the other coil, an e.m.f. is induced in that coil by the changing field. There is no electrical connection between the two coils and the second e.m.f. is said to be 'mutually-induced'.

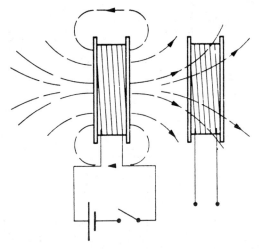

Figure 25.2.

Where a change of current in one circuit results in an e.m.f. being induced in another circuit, the two circuits are said to possess 'mutual-inductance'.

If the direct current source is replaced by an alternating voltage, an alternating field is set up. This causes an alternating voltage to be induced in the second coil having the same frequency as the supply. The double-wound transformer operates on this principle. The coil connected to the supply is the primary, and that in which an e.m.f. is mutually-induced is the secondary.

Transformers can be used to decrease or increase the supply voltage. Transmission is usually high voltage and this is reduced to working voltages by step-down transformers. The generated voltage of a power station is increased in the opposite manner using a step-up transformer. The amount of increase or decrease depends upon the number of turns on the primary and secondary coils. For example, if the primary has 2,000 turns and the secondary 1,000 turns the secondary voltage will be half the applied voltage.

It must be noted that the transformer operates on ac only. For an e.m.f. to be continually induced in the secondary the field must be continually varying. This is possible if the supply voltage is alternating.

Transformer construction
The coils of the transformer are wound on a closed iron core. To minimise the losses in the core due to the alternating current, it is made up of laminations (thin steel strips lightly insulated from each other). The basic diagram of a transformer is shown in Figure 25.3 in which each

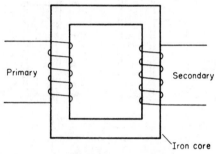

Figure 25.3.

coil occupies one side of a rectangular core. In practice, the coils are either wound over each other (concentric) or wound in sandwich formation. Two types of core formation are commonly used: the core type using a single magnetic circuit or the shell type with a double magnetic circuit. Winding and core arrangements are shown in Figure 25.4.

In order to compensate for volt drop in cables and variations in supply voltage the secondary is usually tapped. These tappings are arranged to give voltages above and below the nominal secondary voltages. Typical values of tapping give $\pm 2\frac{1}{2}$ per cent and ± 5 per cent of the nominal voltage. The tappings are connected via removable links on the secondary.

When the transformer is supplying a load, heat is generated in the windings and core. Since the rating of any electrical machine is governed by the heat which the insulation can withstand, cooling is necessary. The unit is housed in a metal container and cooled by one of the following methods:

Air-cooled
The casing is fitted with a perforated base and louvred sides. Air circulates through the unit by convection. This construction is suitable for dry dust-free conditions and is usually confined to small transformers of about 5 kVA (kiloVoltAmp) or less.

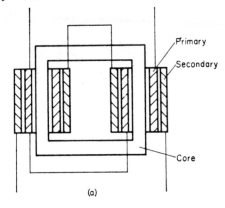

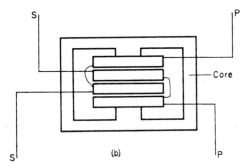

Figure 25.4. (a) Core-type transformer: (b) Shell-
type transformer.

Oil-filled

The transformer is housed in a casing containing insulating oil which completely covers core and windings. The oil acts as a cooling and insulating medium. Pipes are fitted around the casing and the air passing over these pipes carries away the heat that has been transferred from the core through the oil. Figure 25.5 gives the construction of a typical oil-filled transformer. This is the most common construction.

Water-cooled

Very large units are sometimes water-cooled. A coil of copper piping is included in a tubeless oil-filled transformer. The cooling water is pumped through this coil. A fault in the coil could cause the water to mix with the oil. This is overcome by keeping the oil pressure higher than the water pressure.

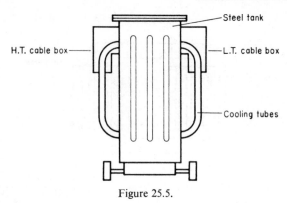

Figure 25.5.

Air-blast cooling

Clean filtered air is forced by a fan through ducts in the core and winding. The method is inferior to oil-cooling as the good insulating qualities of the oil is lost. They are only used in special instances.

Auto-transformers

The transformers previously described are all of the double-wound variety. This means that the primary and secondary, although mounted on the same core, are electrically separate.

The auto-transformer has a single winding only. This winding is tapped at points along the length, and the secondary voltage is obtained by connecting between one end of the winding and a suitable tapping. Figure 25.6 shows the diagrammatic representation of an auto-transformer, and it also explains how the secondary voltage is calculated from the transformer winding turns.

The application of auto-transformers is limited, usually being confined to motor starting. A major disadvantage of the unit is that there is a direct electrical connection between the primary and secondary

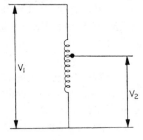

Figure 25.6.

voltages. This could result in the primary voltage appearing at the secondary terminals of the transformer due to an internal fault, and, if the transformer is a step-down type, dangerous conditions could then exist. The transformer can be used to increase (step-up) or decrease (step-down) the voltage, but normally has a ratio of less than 2 to 1.

The construction is generally the same as the double-wound transformer with the same type of cooling. Regulations covering the use of this type of transformer are summarised at the end of this chapter.

Three-Phase transformers
The majority of transformers used in industry and distribution are designed for three-phase operation. In the double-wound type, three separate transformers are wound on a common core. The primary windings are connected in either star or delta, and the secondaries connected likewise. In distribution transformers the secondary is commonly connected in star. This has the advantage of providing a fourth, or neutral, connection and enables the transformer to give *two* values of secondary voltage. Connecting between any of the three winding ends gives full line voltage, and connecting between the neutral and any winding end gives 58 per cent of the line voltage. A typical distribution transformer gives 415V between the lines; connecting between line and neutral gives 58 per cent of this which is 240V. The transformer can therefore be used to supply three-phase loads, such as motors, at 415V, and lighting and heating loads at 240V.

The construction is the same as that previously described, but because of the larger sizes used in three-phase work, cooling is usually by oil. Three-phase auto-transformers are also available, and are mainly used for motor starting.

Applications
Transformers are used in every branch of the electrical industry, ranging from domestic bell transformers to the large three-phase units used in generating stations. Single-phase, air-cooled types are used for bell circuits, radio receivers and discharge lighting. Isolated farmhouses and small housing schemes are usually fed from pole-mounted, single-phase, oil-cooled units. These latter transformers step-down from the rural distribution voltage of, say, 6600V, to the required 240V. Special high-voltage fuses are used to protect the transformer against faults.

The three-phase transformer is used widely in generation and distribution. They step up the generated voltage to the 132kV or 275kV required for transmission and are used to reduce this high voltage to that for normal local distribution. The majority of factories are supplied at 11kV or 6·6kV and have their own transformers to give the required voltages for motors, heating and lighting.

Auto-transformers are used in some types of discharge lighting, but the three-phase types are usually confined to motor starting.

Installation and maintenance

The general precautions to be observed when installing transformers are the same as those for all electrical machinery. The apparatus must be installed to allow its safe operation and protected as far as possible from accidental damage.

Particular care must be taken in the installation of oil-filled transformers to minimise the danger of fire and explosion. A transformer containing more than 100 litres should be housed in a room of fireproof construction, e.g. brick. Where the quantity of oil exceeds 250 litres, a rubble-filled pit should be provided into which leaking or burning oil can drain. The room should have ventilators to the outside. In the case where several large units occupy the same room, blast walls are built between the transformers. Normally, the large transformers used in industry and distribution work, are located outdoors in an annexe to the main substation.

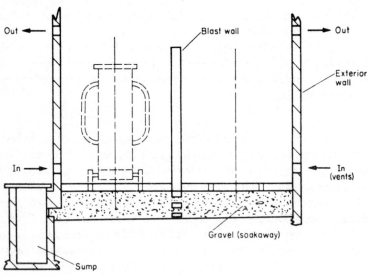

Figure 25.7. Site installation of an oil-filled transformer.

All transformers should be checked at regular intervals for soundness of insulation, using a Megger to check both primary and secondary windings. The oil in oil-cooled units tends to become slightly acid in use and lose some of its insulating qualities. Samples of the oil are drawn off, say every six months, and analysed for acid content and insulation strength. General maintenance includes ensuring that the level of oil is maintained, visual inspection of cable terminations etc., and in the case of air-cooled units, checks for obstruction of ventilation louvres.

Regulations

The following is a brief summary of the regulations relating to electrical equipment. Reference should be made to the current edition of the IEE Regulations and to the various British Standards Codes of Practice which indicate accepted methods of maintenance. Some of these Codes are listed on page 19, Chapter 3 of this book.

1. Chokes and transformers, classified as fixed apparatus, exceeding 60 watts capacity are to be adequately ventilated and enclosed in a proper container or so mounted to minimise fire risks. In the latter case, the mounting must be such that no wood, except hardwood, or combustible material is within 300 mm measured vertically above the apparatus or 150 mm in any other direction from it. Combustible material protected by asbestos or other fireproof material is accepted. The housing of a choke in a discharge-lamp fitting also must comply with the regulations.

2. A step-up transformer, forming part of a consumer's installation, must be provided with a multi-pole, linked switch to completely isolate the transformer from the supply.

3. Transformers containing more than 100 litres of oil must have facilities for draining away any excess oil. The oil must be prevented from leaking into parts of the building. This regulation is met by mounting the transformer over gravel-filled pits.

4. Buildings housing oil-filled transformers should be of fireproof construction. If the quantity of oil is more than 200 litres, the building should be ventilated to the outside.

5. Auto-transformers shall not be supplied from the mains where the voltage exceeds 250V. Exceptions are made if the transformer is used for motor-starting, or with a capacitor for improving power factor. The auto-transformer must be installed beside the capacitor.

Portable appliances, socket-outlets and extra-low voltage apparatus shall not be supplied from an auto-transformer. It should be noted that portable appliances include toys, model railways, etc.

26

Motors and
control gear

General

Motors are the most common and best known of electrical machines. They are classified in two groups depending on whether they are suitable for use on direct-current or alternating-current systems. There is a third motor known as the Universal which can be used on dc or ac and these are used to drive small appliances such as vacuum cleaners, food mixers, etc. Each of the two main groups is subdivided giving several types of motor, each type having its own particular application.

More ac motors are in use than dc motors, and this is due to two factors. The majority of our supply systems are alternating current and allied to this is the fact that the ac motor is simpler and cheaper than its dc counterpart. Where variable speed over a wide range is required such as cranes, locomotives, etc, use is made of the direct-current motor, since the basic ac motor is essentially a single speed machine. This means that where the supply is alternating it is necessary to instal rectifying equipment which adds to the cost of the installation.

The Regulations require that motors above 0·37 kW require control gear which incorporates a device to prevent self-starting and protection against overcurrent. Each type of motor requires its own particular form of control gear and this will be discussed at relevant points throughout the chapter.

Motor enclosures

Motors require some protection, both to prevent the user coming into contact with live and rotating parts and to protect the motor from damage due to dampness, dust, etc. All motors consist of a stationary part called either the yoke (dc) or the stator (ac) and a rotating part referred to as the armature (dc) or the rotor (ac).

In the case of very large motors driving machinery in a dry dust-free atmosphere the rotating part is often supported on pedestal bearings. The motor is then fenced off to prevent anyone from coming into contact with the conductors or moving parts. It is obvious that this method is only permissible in ideal atmospheric conditions and where no work is carried out near the motor. In other situations a more robust construction is required and details of various enclosures are given below.

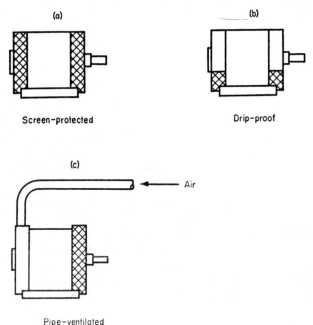

Figure 26.1. Types of motor enclosures.

(*a*) *Screen-protected*. The motor is enclosed in a steel casing and the shaft runs on bearings housed in the end covers of the casing. The end covers are fitted with metallic screens and an internal fan drives cooling air through the motor. This type of motor is used in dry, dust-free conditions and is the most common enclosure in use.

(*b*) *Drip-proof*. This enclosure is similar to the screen-protected type, but the screens are fitted with angled cowls. These cowls prevent any dripping water from entering the machine. The motor can be used in damp conditions but it must be remembered that the enclosure is not waterproof. Again, the enclosure is not suitable for dusty atmospheres as particles of dust blocking the screens could cause overheating.

(*c*) *Pipe-ventilated*. The air which is circulated through the motor by the internal fan is brought from outside the building in which the motor is located. This type of enclosure is useful when the motor is located in very dusty situations such as woodworking shops.

(*d*) *Totally-enclosed.* There are no air inlets in this type of enclosure, the cooling being by means of radiation from the surface of the motor. An internal fan is fitted and the casing is ribbed to increase the cooling surface. The motor runs hotter than any of those previously described and a high-class insulation material such as fibre glass is often used to withstand the higher temperatures. Dirt and moisture are excluded, and such an enclosure is suitable for use in corrosive atmospheres. Care must be taken that the ribs do not become clogged up or overheating will occur.

(*e*) *Flameproof.* In construction these are similar to the totally enclosed machines but are physically larger and more robust. The end covers are bolted to the main casing, which is fairly thick metal; and special seals are fitted at the shaft bearings. The construction has to be sufficiently strong to withstand an internal explosion without allowing the passage of a dangerous flame which could ignite the external atmosphere. These motors are used in the mining and oil industries or any situation where explosive atmospheres are encountered. Great care must be exercised in the maintenance of these machines.

In addition to the above general types, water-cooled motors are also available. Water is pumped through ducts in the stationary part and through the hollow shaft and along ducts in the rotating part. Several firms also make submersible motors which are completely watertight. These are used for driving pumps in wells and boreholes.

DC motors

This book does not go deeply into the theory of the electric motor: the following notes give only brief details of the action.

The motor basically consists of a set of coils wound on a laminated-steel core or armature. The armature is located on a shaft running on bearings and is free to rotate between the poles of a magnet. When current is passed through the coils a force is produced causing the armature to turn. The magnitude of the force depends on (*a*) field strength, (*b*) value of current and (*c*) length of coil.

In theory the coils will only turn until they lie at right-angles to the magnetic field, but by reversing the current when the coils are midway between the poles continuous rotation is produced. This reversal is achieved by using a commutator, which is a set of copper segments insulated from each other and from the motor shaft. The ends of each coil are connected to segments of the commutator and current led in and out of the coils by brushes running on the commutator. Figure 26.2 shows the general construction of a small direct-current motor.

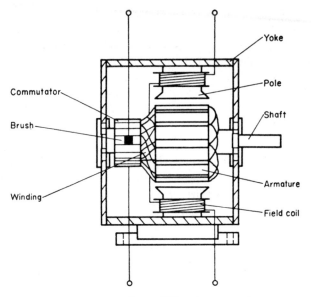

Figure 26.2.

The coils are laid in slots on the armature and are held in place by wooden or fibre wedges. In the simple motor shown the coils are connected in series and the ends of each coil brought out to segments on the commutator. The carbon brushes are held in spring-loaded boxes and run on the commutator surface. In very large machines several brushes, connected in parallel, may be required to carry the heavy currents involved. Ball and roller bearings, carried in the main casing or in external pedestals, support the shaft. Sleeve bearings, in which the shaft turns in cast-steel, white-metal lined, oil-lubricated sleeves, are often used in large machines. The shaft is machined and a keyway provided to allow the fitting of pulleys, couplings or gearing.

The stationary part, or yoke, carries the magnet poles and is also part of the magnetic circuit. In practice electromagnets are used and the strength of the magnetic field is controlled by varying the current flowing in the magnetising coils. It will be seen later that by varying the field current, which in a large machine is only a few amperes, the speed of the motor can be varied. The pole pieces and yoke are usually cast-steel and the poles are bolted to the yoke. Owing to the sudden reversal of the current in the armature and the distortion of the field due to armature current (armature reaction) severe arcing at the brushes can occur. To minimise this sparking small electromagnets known as interpoles are often fitted between the main poles. The coils on these interpoles are connected in series with the armature and carry the armature current.

To assist cooling an internal fan is fitted in the motor shown.

There are several types of direct-current motors, classified according to the method used to connect the armature and field systems to the supply. The types in general use are (*a*) shunt, (*b*) series and (*c*) compound.

Shunt motor

In this motor the field coils are connected in parallel (shunted) across the armature. Usually a variable resistance is included in the field circuit to control the field current and therefore the speed of the motor. The circuit diagram of a shunt motor is shown in Figure 26.3.

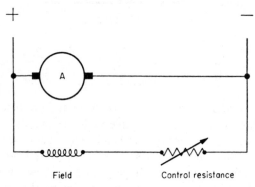

Figure 26.3.

The speed of the shunt motor falls slightly as the load it is driving increases, but can be taken as fairly constant over the range between no load and full load: above full load the speed drops rapidly. The motor is thus suitable for driving machines such as pumps, lathes, etc., where sudden overloads do not occur. This is the most common dc motor.

Starting and speed control. All except the smallest of motors require some type of starter to prevent heavy currents being drawn from the supply on starting. For dc machines the basic requirement is a graded resistance in series with the armature to limit the starting current. The resistance is cut out of the circuit in steps as the motor speeds up. To comply with the regulations, however, means to prevent self-starting after failure of supply, and overcurrent protection is added to the basic starter. The starter is required because on switching on the motor the current would be limited only by the armature resistance (about 1 ohm), only if no series resistance is used. As the conductors begin to turn an e.m.f. is generated in them in a direction opposite to the applied voltage. This e.m.f. is known as the 'back e.m.f.' and the voltage acting on the armature at any time is the difference between the applied voltage and the back e.m.f. The value of the back e.m.f. of a standard dc machine on normal duty is only a few volts less than the applied voltage. As an example the back e.m.f. of a 400V motor may be 390V, the voltage acting on the armature being $400V - 390V = 10V$. The small resultant voltage is sufficient to circulate the armature current required to develop the correct horse-

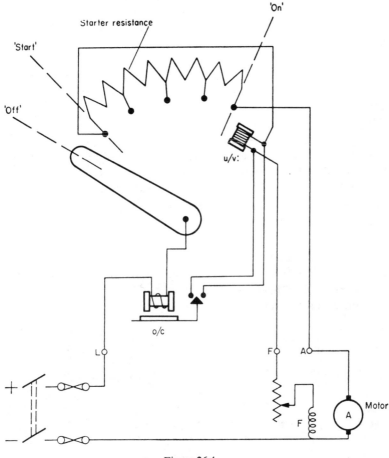

Figure 26.4.

power. The starter is therefore cut out in steps as the back e.m.f. builds up with the increase in motor speed.

The faceplate starter is suitable for use with a shunt motor. The device to prevent self-starting is called the under-voltage coil and is an electromagnet through which the field current passes. The starter handle is attracted and held in place by this electromagnet when the starter is in the fully ON position. Should the supply fail, or the field circuit become broken, the coil is de-energised and the handle is pulled, by the spring, back to the OFF position.

Overcurrent protection is provided by a second electromagnet, the coil of which carries the motor current. If the current becomes excessive the trip-bar is attacted to the electromagnet and the contacts close, short-circuiting the under-voltage coil. The starter handle is again released and returned to the OFF position.

It should be noted that the action of the starter is such that the field circuit is completed as the handle moves to the first stud of the starter. If the supply to the armature is maintained without the field circuit being complete, the motor, if running, will tend to speed up and may reach dangerously high speeds.

The speed of the motor can be changed by either varying the armature voltage or the field current. Invariably variation of the field current is used as only a few amperes flow in this circuit, whereas the armature current may be very high and any control gear would have to carry this large current, if armature voltage control were used. Reducing field current increases the speed.

Series motor

The armature and the field are connected in series as shown in Figure 26.5. The field coils carry the armature current and consist of only a few turns of heavy-gauge conductor. When the motor is on light load the motor runs at a high speed due to the weak magnetic field. As

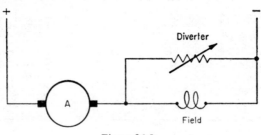

Figure 26.5.

more load is applied to the motor the current rises, the field strengthens and the speed drops rapidly. At the lower speed the motor is able to accelerate very heavy loads and is therefore very suitable for driving cranes, electric locomotives, etc.

This type of motor should always be directly connected to the load by means of gearing or bolted couplings. Belt drives should not be used as a broken belt causes the motor to go on 'no-load' and the motor speed may increase to a dangerous level.

Starting and speed control. The starter in this case is a high-rated resistance in series with the armature. Overcurrent and under-voltage protection is provided as in the shunt-motor starter. Speed control is effected in traction and hoist work by using the heavily-rated starting resistance as a controller to vary the voltage across the motor terminals. A degree of speed control is achieved by by-passing some of the field current through a 'diverter' shunted across the field.

Compound motor

This motor has both shunt and series fields as shown in Figure 26.6. The fields are connected so that they may assist each other, strengthening the field with increased load, or oppose each other. Both field coil systems are wound on the same motor poles; the degree of 'compounding' causes the motor speed to vary with load in a particular way. The speed can be made to rise, remain constant or fall with increase in load. If the speed rises slightly, or stays constant with increase in load, the motor is

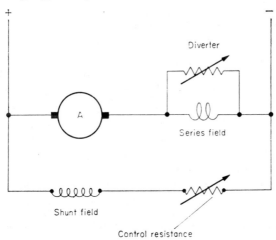

Figure 26.6.

said to be 'differentially' compounded. A reduction in speed is obtained from a 'cumulative' motor where the fields assist each other.

The differential motor is useful for driving machines where a constant speed is necessary under varying load conditions, such as a processing machine. The slightly series characteristic of the cumulative motor makes it suitable for driving machinery where sudden heavy loads are imposed for short periods. An example of this is a strip rolling mill where ingots of steel are passed through rollers to be rolled into steel plates. This motor does not suffer from the disadvantage of series motors that it will overspeed if the load is removed.

Starting and speed control. The compound motor is started using the same faceplate starter as already described for the shunt motor. Speed control is effected by a variable resistor in the low-current shunt field circuit.

Reversal of dc motors

A dc motor is reversed by changing over the connections to *either* the armature or field circuits. Reversal of the mains supply leads will cause the motor to rotate in the same direction as before. Where a reversing switch is used this is always connected in the shunt field circuit in shunt and compound motors. The reason for this is that these fields carry very small currents relative to the armature.

AC motors

Most of the motors used in industry and the home are of the alternating-current pattern. Industrial motors are usually designed for three-phase operation and the domestic motors such as used on washing machines, etc. are designed for the single phase systems used in the home. The size of single phase motors used in industry seldom exceed 7 kW.

Three-phase motors are fairly simple in construction and have the advantage of being self-starting. The single-phase motor is not self-starting and requires to be wound in a special way and fitted with special starting arrangements. This makes it much more expensive and larger than a three-phase motor of equivalent rating.

One minor disadvantage of the ac motor is that it runs at a fixed speed in standard motors. The speed depends on the number of poles and the frequency of the supply system. The relationship between the speed, poles and frequency is given by the formula $N = \dfrac{60f}{p}$ where 'N' is the approximate speed in revs/min., 'f' the frequency in Hertz and 'p'

the pairs of poles in the motor. As an example the speed of a 4-pole motor operating on a 50 Hz supply system is $\dfrac{60 \times 50}{2} = 1500$ rev/min approximately.

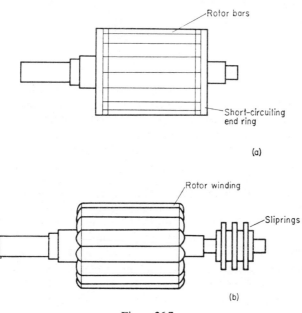

Figure 26.7.

The speed of some industrial motors is changed by 'pole-changing' in which the motor windings are switched to give different numbers of poles. It should be noted that this system only gives different values of fixed speeds. Variable-speed commutator ac motors are also available but these are out of the scope of this chapter.

The motor consists of a fixed portion, the stator, which is built up of laminated steel plates. Laminations are used, rather than a solid core, to minimise the eddy current losses associated with ac magnetic circuits. The stator carries the field winding which is distributed in slots around the core, and the winding is retained in the slots by wooden or fibre wedges.

The rotating part, rotor, is made up of laminated plates, mounted on the motor shaft, and the rotor can be of the squirrel-cage or wound-rotor design. Both types of rotor are shown in Figure 26.7. A squirrel-cage rotor has lightly-insulated copper or alloy bars laid in slots around the rotor. These bars are short-circuited at each end of the rotor. This is the most common type of rotor construction in use. In the wound rotor, coils are laid in the rotor slots and the coil ends connected to slip-rings on the motor shaft. Carbon, or composition, brushes running on the slip-rings allow the rotor to be connected to an external, variable, resistance for starting purposes.

Three-phase motors
The basic principle of these motors is as follows: Three sets of stator coils are uniformly distributed around the stator core and connected in either star or delta. The rotor can be of the squirrel-cage or wound-rotor type and if the latter is used the three rotor windings are usually in star and brought out to three slip-rings.

When the stator is connected to the ac supply, the field produced rotates around the stator at a speed given by the formula previously quoted $\left(N = \dfrac{60f}{p} \right)$ The machine behaves like a transformer where the stator is the primary and the rotor the secondary. The ac applied to the stator causes an alternating voltage to be induced in the rotor and a current therefore circulates in the short-circuited rotor winding. The rotor conductors now carry current in a magnetic field and a force is produced causing the rotor to turn in the same direction as the rotating field. The speed of rotation is slightly less, however, than the speed of the field which rotates at 'synchronous' speed. The difference in the two speeds is the 'slip' of the motor and is of the order of 5 per cent of the synchronous speed of standard motors at full load.

It should be noted that there is no electrical connection between the stator and rotor. The action is due to rotor induced voltages and the motor described is often referred to as an 'induction motor'.

Applications
The speed characteristic of an induction motor is similar to that of the shunt motor, the speed being fairly constant up to full load. Squirrel-cage motors are used for driving machines having a steady load such as workshop lathes, centrifugal pumps, air compressors and other machines which can be started on light load. Where the supply system can withstand the starting current surges, squirrel-cage motors of up to 2,500 hp and 11,000V can be used. Normally, however, the rating of squirrel-cage motors rarely exceeds 75 kW where the industrial consumer is supplied from a medium sized substation, at 3,300V and 415V.

Slip-ring motors are used where heavy machinery has to accelerate or where the machine is started against full load. Examples of these conditions are hoists, mining haulages, large ventilating fans, etc.

Control gear

Small squirrel-cage motors, starting on light load, can be switched direct-on-line, but to comply with the IEE Regulations the starters are fitted with isolator, and overcurrent and under-voltage protection. On starting, the motor, if switched direct, takes a current of up to seven times its normal full load current, which decreases as full speed is reached. The supply authorities require that the magnitude and duration of these starting surges be limited and use is often made of special starting gear.

Starting equipment for wound-rotor (slip-ring) motors comprises two parts: a stator switch, which contains the protective devices, and a graded rotor resistance which is cut out as the motor speeds up.

(*a*) *Direct-on-line starting*. This can be a hand operated switch, but more use is now being made of the push-button operated contactor starter. Figure 26.8 shows the circuit diagram of a simple contactor starter.

The start button, when pressed, completes the contactor circuit and closes the three-phase switch establishing the supply to the stator. The motor current passes through the overload coils and if the current becomes excessive the overcurrent solenoid operates the trip coil and breaks the contactor circuit. Undervoltage protection is inherent, as failure of the supply causes the contactor, which is operated from the supply, to drop out. The stop-button is wired in series with the overload contacts to break the circuit. An isolating switch is mounted in, or near to, the starter so that the circuit can be isolated, and if necessary this can be locked off for safety during motor maintenance.

If a hand-operated starter is used the overcurrent device actuates the mechanical linkage of the starter. In this case, the undervoltage coil is energised from two of the supply terminals and also actuates the mechanical linkage if the supply fails.

(*b*) *Star-delta starting*. This method of starting requires that the two ends of each of the three stator coils is brought out to a six-terminal box. The motor windings are connected in star at starting which means that each winding receives approximately 58 per cent of the supply voltage. When the motor approaches its full speed the windings are switched over to give a delta connection in the full supply voltage applied to each winding and the motor develops its full horsepower. The current taken

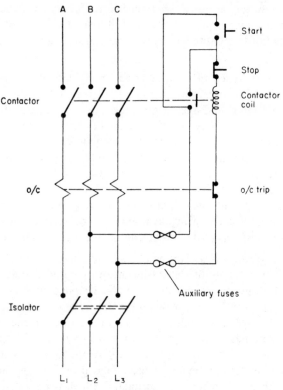

Figure 26.8.

when the windings are in star is reduced, but the horsepower the motor can develop is only one-third that obtainable in delta. The method is only suitable, therefore, where the load being driven is reasonably easy to accelerate. Figure 26.9 shows the circuit diagram of a typical hand-operated star-delta starter. The drawing shows the devices required to comply with the regulations. Greater use is being made of automatic push-button starters using contactors. A timing device is used so that one contactor closes at starting for the star connection then after a pre-set time a second contactor closes to connect the windings in delta.

(*c*) *Auto-transformer starting*. Where the reduction in horsepower output caused by the star-delta method of starting is too great use is made of the auto-transformer starter. In this case the motor is supplied from an auto-transformer which is tapped to give a reduced voltage on starting and then switched to give full voltage as the motor runs up to speed.

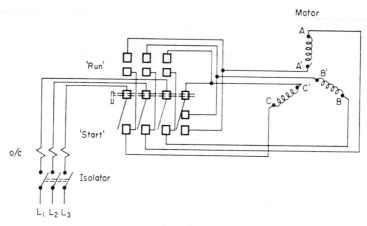

Figure 26.9.

Typical values of starting voltage are 70 per cent to 80 per cent supply voltage. Figure 26.10 gives circuit details of the starter.

(*d*) *Rotor starter.* This method is used for slip-ring motors and the circuit is given in Figure 26.11. To avoid very high surges on starting the resistance must be 'all-in' before closing the stator switch. A mechanical interlock is usually fitted so that when the stator switch is switched off, the rotor-resistance arm is automatically returned to the 'all-in' position, ready for the next time the motor is to be started. The

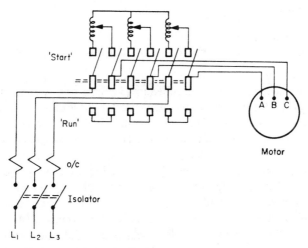

Figure 26.10.

rotor resistance gives some degree of speed control, on load, by switching out portions of the resistance only to give a speed lower than full speed. The resistance used for speed control must be very heavily rated to dissipate the heat generated by the rotor currents, and even then can only be used for short periods of time.

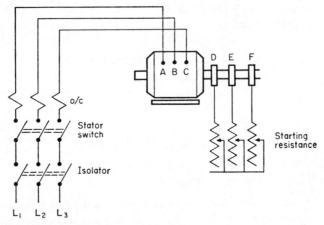

Figure 26.11.

Again, an automatic starter is available, using timers and contactors to cut out sections of the resistance at pre-determined intervals.

Reversal of three-phase motors is obtained by changing over any two of the supply leads.

Single-phase motors
There are two main types, the induction motor and commutator motor.

Induction motor
If a single-phase ac voltage is applied to a squirrel-cage induction motor with only one stator winding, the magnetic field produced is not rotating but merely pulsates. The rotor will not turn, but if turned mechanically will start to rotate on its own accord after a certain speed is reached; it will, however, rotate in either direction. In order to produce an equivalent rotating field a second winding, the starting winding, is also wound on the stator at 90° to the main winding. This winding produces a field which is out of phase with that due to the main winding and several methods are employed to obtain the phase displacement. Motors constructed in this fashion are called 'split-phase' induction motors. To minimise losses, and keep the power factor as high as possible, a centri-

fugal switch is fitted on the motor shaft which open-circuits the starting winding when the motor attains about 80 per cent of full speed.

(*a*) *Inductor-start split-phase motor.* The starting winding in this motor is of higher inductance than the main winding so causing phase displacement at starting. To obtain even greater displacement, and better starting characteristics with reduced current, a choke may be placed in series with the starting winding. Figure 26.12 shows the connection arrangements at start and when running.

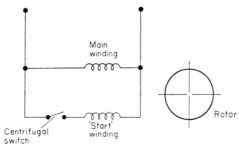

Figure 26.12.

The motor is reversed by changing over the connections to either the main or auxiliary windings.

(*b*) *Capacitor-start split-phase motor.* If a capacitor is connected in series with the starting winding, instead of a choke as described above, an even greater phase displacement is obtained. A common variation of this is to have both the main and starting windings rated for continuous running and use two capacitors. When the motor runs up to speed the centrifugal switch cuts out one capacitor and leaves the other permanently connected to give power factor correction. Figure 26.13 shows the arrangement for a capacitor-start, capacitor-run motor.

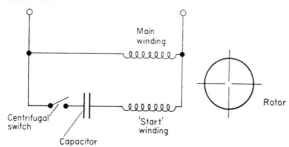

Figure 26.13.

Reversal is obtained as before, by changing connections to either main or auxiliary windings.

Both motors have speed characteristics similar to dc shunt motors and are used for driving workshop machinery and the larger domestic equipment such as refrigerator drives.

Starting is usually direct-on-line for the smaller machines or a series resistance can be used in the stator circuit and cut out as the motor speeds up. The majority of split-phase motors have squirrel-cage rotors, but wound rotor types are available for heavier starting duties. The wound rotor can be exactly as that used in a three-phase machine, and the three slip-rings are connected to a starting resistance.

Commutator motor

The motors described here are single-speed machines, although it must be remembered that special variable-speed commutator motors for use on ac are available.

(*a*) *Single-phase series motor*. As stated in the section on direct-current motors, reversal of the dc mains causes the motor to run in the same direction. If a dc motor were used on an alternating supply the current in the armature and field would reverse simultaneously. This would mean that the motor would run in one direction, and series motors are used on ac. These are often referred to as Universal motors since they are suitable for ac/dc working. When used on ac there is some trouble with commutation which, in a large motor, could give rise to severe arcing at the brushes. The motors are used only in small sizes and find their application in various domestic appliances. The speed characteristic is that of the dc series machine, the speed falling with increased mechanical load. They are used to drive articles having a fairly constant load such as hand tools, vacuum cleaners, etc. To minimise losses when used on ac the stator (yoke) and rotor (armature) are built up of laminated steel, but otherwise the machine is the same as a dc motor.

Direct-on starting is usual for these small motors. Reversal of rotation is obtained as in the dc case, that is, reverse either field or armature winding.

(*b*) *Single-phase repulsion motor*. This consists of a single-winding on the stator and a commutator armature. Short-circuited sets of brushes run on the commutator surface. When the motor is switched on, the induced e.m.f. in the armature coils causes a current to flow in the armature. In order that rotation should take place, the brushes have to be moved round the commutator until their axis is at an angle to the stator field as in Figure 26.14. Interaction between the fields due to stator and armature currents then produce rotation. The speed of the repulsion motor behaves as that of a series motor, but the motor is capable of starting fairly heavy loads, and is available in much larger sizes than the series motor. The most suitable application is for constant speed heavy drives. Starting may be direct-on or by means of a series stator resistance. The direction of rotation is reversed by moving the brushes round the surface of the commutator.

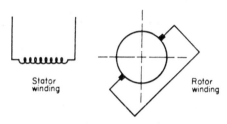

Figure 26.14.

(*c*) *Single-phase, repulsion-start, induction motor.* The construction is practically the same as the basic repulsion motor. The motor is started as before, but as the speed increases a centrifugal mechanism causes a ring to short-circuit the whole of the commutator and lift the brushes clear. The armature is thus converted into a squirrel-cage rotor and the machine runs as an induction motor. This type of machine is only used in small sizes and has the same applications as the induction motor, but with better starting characteristics. Reversal of rotation is again achieved by shifting the position of the brushes.

(*d*) *Single-phase, repulsion-induction motor.* This machine has similar characteristics and applications as the repulsion-start, induction motor. The armature has a double winding, an outer commutator winding, and a squirrel-cage inner winding in the same slots. The motor starts as a repulsion type, but runs as an induction motor. There is no need for a short-circuiting and brush-lifting device; the motor has a fairly good power factor. Starting and reversal of rotation is as before.

Single-phase synchronous motor
These motors are used to drive clocks and other accurate timing devices. The speed is dependent on supply frequency which is maintained normally at a very accurate figure by the supply authorities. One form of synchronous motor is that known as the 'shaded-pole' motor. This

motor consists of a horseshoe-shaped, laminated iron core on which the magnetising coil is wound. A copper ring is embedded on the face of each of the pole-pieces as shown in Figure 26.15. The effect of this ring is to split the magnetic field into two parts which are out of phase with each other and give a rotating field effect. The rotor is simply a slotted rod, and when the motor is switched on the rotor runs at synchronous speed, i.e. the speed of the rotating field. This type of motor is self-starting, but is only suitable for driving light loads such as clocks and gramophone turntables at constant speed.

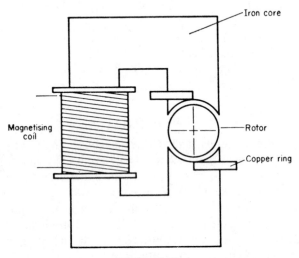

Figure 26.15.

Installation of motors

Serious consideration must be given to the type of motor enclosure used in a particular situation. Before the motor is installed an insulation test should be made. If this is low the motor should be stored in a warm place to drive out any moisture. Another method of drying out is to lock the armature or rotor and pass a current from a low-voltage supply through the windings; but this must be used carefully to avoid over-heating.

Motors are usually mounted on steelwork firmly secured in a concrete base. Where pulley drives are used some method of belt tensioning is required and the motor is fixed on slide rails as shown in Figure 26.16. Adjusting screws bearing on the motor feet between motor and load are used for tensioning. The pulleys on the motor and the driven load must be exactly in line and parallel to each other. This is best checked by lining up the pulleys by a straight edge or line and Figure 26.17 shows the method of alignment.

Pulleys and gearing are retained on the shaft by means of a key. The key should be of the drilled and tapped type which can be easily withdrawn using extracting gear. There should be no movement of the pulley on the shaft with a correctly fitted key, but it is necessary to have a little top clearance between the key and the pulley hub. Keys should never project beyond the end of the motor shaft. Where the motor is fitted on slide rails for adjustment, the cabling should be terminated in such a way that it allows movement of the motor. When armoured cables are used this presents little problem, but conduit installations

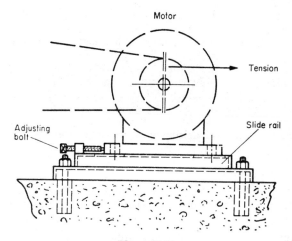

Figure 26.16.

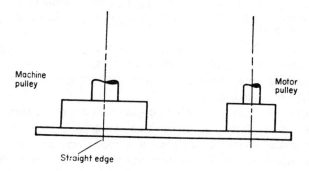

Figure 26.17.

require flexible extensions at the motor. Mineral-insulated cables are installed with a loop of cable at the motor termination point.

When the motor is ready to be put into service a final insulation test of the complete installation should be made. The bearings should be checked to ensure they are charged with a sufficient quantity of grease or oil.

Summary of applications

Motor type	Application
DC shunt	General purpose; constant load, machine drives.
DC series	Starting against heavy load, e.g. traction.
DC compound	
(a) cumulative	Fluctuating heavy loads, e.g. rolling mills.
(b) differential	Constant speed, e.g. processing.
3-Phase squirrel-cage	General purpose. Light-load starting, e.g. machine drives.
3-Phase, slip-ring	Accelerating heavy loads, e.g. hoists.
1-Phase, induction	As 3-phase, squirrel-cage, but limited to about 3kW.
Repulsion	As induction, but capable of starting heavier loads.
Universal	Small domestic appliances; hand tools, usually about 75W.
1-Phase, synchronous	Timing equipment; Record-players and tape-recorders.

Summary of Regulations

The Regulations cover both dc and ac machines and lay down requirements regarding housing and construction of machines, provision of control gear, rating of cables and permissible volt drop in motor circuits.

The Regulations are summarised as follows:

1. (Reg. B23) The volt drop between consumer's terminals and motor terminals must not exceed 2·5 per cent of the declared or nominal voltage.
2. (Regs. C8–C10) Where there is a danger of fire or explosion, the motor and its control gear shall be of flameproof construction. Alternatively, in dust-laden situations, the motor enclosure shall be of a type to exclude dust.
3. Motors used in situations where the temperature of the surrounding atmosphere is such that overheating could occur, shall be downrated or the insulation shall be of a special material. Means for forced ventilation, or pipe-ventilated enclosures, can be used as an alternative.

4. (Reg. A64) Every electrical motor shall be provided with efficient means of starting and stopping. This control shall be within easy reach of the operator. In addition, motors of more than 0·37 kW shall be provided with control gear with the following devices: (a) means to prevent self-starting in event of restoration of supply after failure, (b) protection against overcurrent in the motor and in the cables supplying the motor and (c) means of isolating the motor and all its associated apparatus.

In modern factories use is made of group control whereby several motors are automatically started from a group of contactors on a single board. It is useful to have a stop-button close to each motor position. Normally this stop-button is of the lockable type and, in addition, an isolator can be provided at the motor position. The main isolator on the board should be lockable to comply.

5. (Reg. A67) The final subcircuit, supplying a motor, shall be protected by fuses, the current rating of these being not greater than that of the cable, or by a circuit-breaker whose setting is not greater than twice the cable rating (A68). Where fuses are used in conjunction with control gear as specified in item (4) then their rating can be twice that of the cable. The rating of cables in rotor circuits, or commutator circuits, of induction motors is also included (A69). And these cables must be suitable for starting and running conditions.

27

Capacitors and power factor

Capacitors are used in installation work mainly for power factor correction, starting of single-phase motors and anti-interference devices. A brief description of the action of capacitors and the effect and importance of power factor is given here. For fuller treatment of both subjects, a good electrical science book should be consulted.

Capacitor action
The basic capacitor consists of two plates, separated from each other by a layer of air or an insulating material known as the 'dielectric'. If this arrangement is connected to a dc source, electrons leave one plate and travel round the circuit to the other plate. During this period a current flows for a short time before falling to zero. An excess of electrons on one plate and a deficiency on the other causes a potential difference (voltage) to exist between the plates. When this voltage is equal and opposite to the applied voltage, the current flow ceases. The capacitor is then 'charged' and will remain so even when the external supply voltage is removed. Connecting the terminals of the capacitor to another circuit will cause a current to flow until the potential difference between the plates falls to zero. The capacitor is then 'discharged' and the electrons rearrange themselves to give this condition. Figure 27.1 shows the action of a capacitor in a dc circuit.

If the capacitor is now connected to an alternating voltage source the conditions are changed. The capacitor charges as the voltage rises from zero during the first quarter cycle, and discharges as the voltage returns to zero during the next quarter cycle. The process is repeated throughout the next half cycle, but in the reverse direction. It can be seen that steady conditions never exist and an alternating current, of the same frequency as the supply, flows. This current, however, leads the applied voltage by 90°, that is by a quarter cycle. During the operation ac electrostatic fields are produced which cause energy to surge to and from the capacitor in alternate quarter cycles.

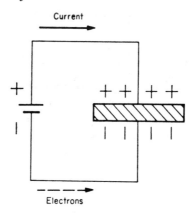

Figure 27.1.

Power factor

Most of the machines used in electrical work are inductive (e.g. motors, transformers). An inductive circuit is one which possesses resistance and inductance. In such circuits the current lags behind the applied voltage by an angle ($\theta°$) which is between zero and 90°. The current has two components, the power component, in phase with the voltage, which does useful work and the 'idle' or reactive component 90° out of phase with the voltage. This idle component is responsible for the setting up of magnetic fields, causing energy to surge to and from the load during alternate quarter cycles. It should be noted that this is similar to the energy in a capacitor. If an inductor and capacitor are connected to the same ac source, the inductor would take in energy, as the capacitor was giving out energy (discharging).

The power component of the current is responsible for doing useful work (e.g. developing the machine output) and is constant for a given load. If the angle between the total current and voltage is reduced, without altering the power component, the total current for a given duty is reduced. The ratio of the power component to total current is the 'power factor' of the circuit and is a figure normally less than 1. Typical values of power factor for an induction motor is 0·8 lag; this means the total current lags the voltage by an angle whose cosine is 0·8. If the total current in this instance is 10A then the power component is $10 \times 0.8 = 8A$. Increasing the power factor to say 0·9 on a load which requires a power component of 8A, reduces the total current to $8 \div 0.9 = 8.88A$. It can be seen that the same work can be done with

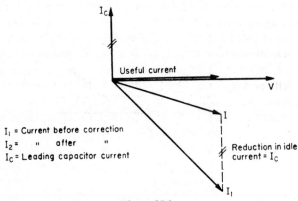

I_1 = Current before correction
I_2 =   "   after   "
I_c = Leading capacitor current

Figure 27.2.

less current by increasing the power factor. Where very large loads are supplied this can mean considerable savings in cable size. In addition, supply authorities impose a charge for poor power-factor; and energy costs are also reduced by improving the power-factor. Figure 27.2 gives the phasor diagram before and after correction.

The most common method of improving the power factor of a machine is by connecting a capacitor in parallel with the machine. The capacitor takes a leading current; this compensates for and reduces the idle component of the machine current. The power factor is therefore increased and the total current reduced. Care must be taken that the capacitor used is not too large or the total current will lead the voltage. This would produce an idle component as before and increase the current. Over-correction of the power factor is sometimes avoided by use of a device to automatically cut out the capacitor when the machine is on light load.

Capacitor construction
Most of the capacitors used in installation work are metal-foil, paper-insulated. The capacitance depends on the area of the plates and on the thickness and the insulating qualities of the dielectric. A common method of increasing the area of the plates is to use two metal-foil strips separated by wax- or oil-impregnated paper. The physical size is kept down by rolling the plates and paper to give a tubular capacitor. The whole assembly is then housed in a plastic container, although higher voltage units and larger sizes are usually housed in a steel casing. In smaller sizes this is the type used for single-phase motor starting and power-factor correction in discharge-lamp circuits.

Where the power factor of large motors is to be improved, the paper capacitor is again used. In order to give the required capacity several capacitors are made up in banks. The banks are placed in a steel casing which is then filled with insulating oil in the same way as a transformer. If the motor, or load, is three-phase, three banks are used and connected in delta or star as required. The three banks of capacitor normally share the same tank.

As mentioned earlier the capacitor remains charged when removed from the supply, and, depending on the quality of insulation, can remain charged for a considerable time. To avoid danger of shock when working on a capacitor discharge resistors are fitted to the larger sizes. These resistors are permanently connected across the capacitor terminals and are of a high value, usually several Megohms. They allow any charge on a capacitor to be safely dissipated after the capacitor is switched off. Care must be taken during capacitor maintenance not to remove these resistors. Before working on a power factor correction capacitor, at least ten minutes should be allowed, after switching off, to allow the capacitor to discharge.

The installation and maintenance of oil-filled capacitors is generally the same as that of transformers. Fire and explosion precautions must be taken and periodic checks on oil samples made. The capacitors used on lamps and motors are connected across the terminals of the apparatus (i.e. in parallel). Correction of an entire system (e.g. of a factory) is sometimes done by installing large capacitors in the main substation. In this case, each capacitor is controlled by a circuit-breaker, with protection against overload of cables and apparatus faults.

Capacitors used in radio work are sometimes of the multi-plate or electrolytic type. The multi-plate construction consists of a set of fixed and a set of moving plates. The capacitance is altered by moving the moving plates in or out of mesh with the fixed plates. The insulation in this case is air. The electrolytic capacitor is similar in appearance to the tubular paper capacitor. The plates are formed by electrolytic action which takes place when the capacitor is connected to the supply. This type of capacitor must connected in the circuit with the correct polarity or damage will result.

A further use of capacitors in installation work is to act as a suppression device on an inductive circuit. The capacitor is connected across the contacts of a switch, such as the make-and-break of a trembler bell. The energy of the spark is then used up in charging the capacitor. In the section on lighting, the diagrams show a capacitor across the starting switch. This acts as a suppressor and prevents radio interference.

Regulations

The general regulations covering the installation and operation of electrical machines and apparatus also apply to capacitors. Precautions must be taken against overheating and fire risk. Every capacitor, except radio interference types, must be provided with means of automatic discharge. Small capacitors are exempt from this requirement. Plugs containing capacitors for radio interference must comply with the requirements of BS613. It is suggested that the capacitor should be fitted to the appliance rather than the plug. (Note: Refer to IEE Regulation C35.)

28

Circuit and wiring diagrams

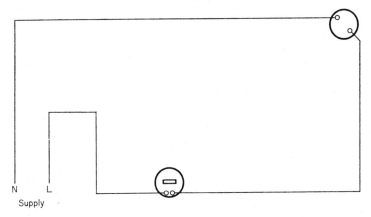

Figure 28.1. Wiring diagram of a lamp controlled by a one-way single-pole switch.

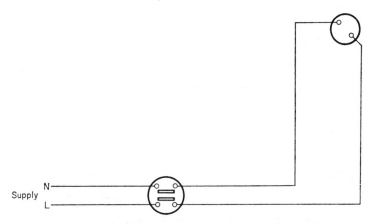

Figure 28.2. Wiring diagram of a lamp controlled by a double-pole switch.

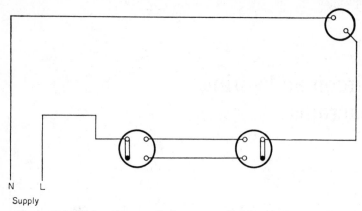

Figure 28.3. Wiring diagram of a lamp controlled by two 2-way switches.

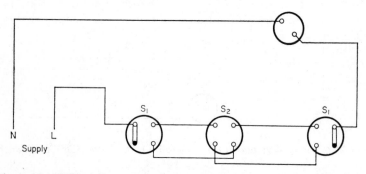

Figure 28.4. Wiring diagram of a lamp controlled by an intermediate switching arrangement with two 2-way switches and one intermediate switch.

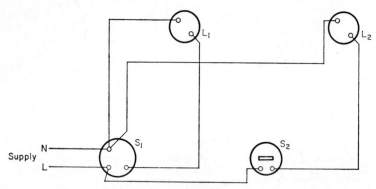

Figure 28.5. Wiring diagram of one lamp controlled by a one-way switch with a loop-in terminal and one lamp controlled by a one-way switch.

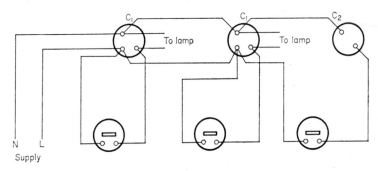

Figure 28.6. Wiring diagram of a lighting circuit using 3-plate ceiling roses and a 2-plate ceiling rose.

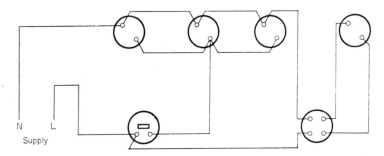

Figure 28.7. Wiring diagram of a lighting circuit with three lamps in parallel controlled by a one-way switch, and one lamp controlled by a double-pole switch.

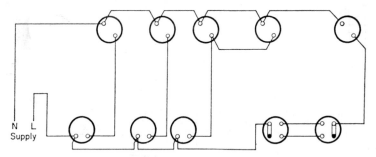

Figure 28.8. Wiring diagram of a typical lighting final subcircuit.

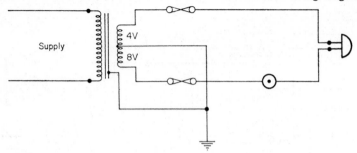

Figure 28.9. Simple single-bell circuit supplied from a bell transformer.

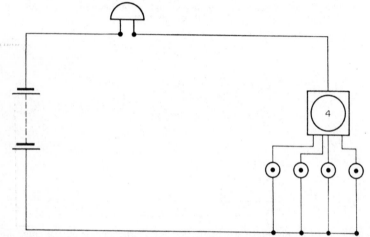

Figure 28.10. Battery-supplied bell circuit with a 4-way bell indicator. Bell operated from any bell push.

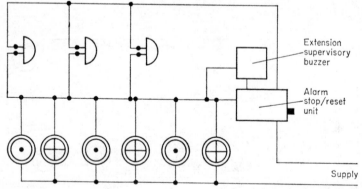

Figure 28.11. Schematic diagram of an open-circuit alarm system with an alarm stop/reset unit.

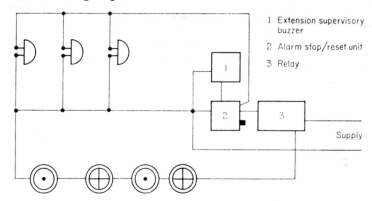

1 Extension supervisory buzzer

2 Alarm stop/reset unit

3 Relay

Supply

Figure 28.12. Schematic diagram of a closed-circuit alarm system with an alarm stop/reset unit.

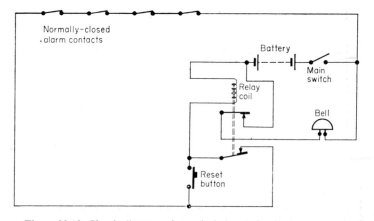

Figure 28.13. Circuit diagram of a typical closed-circuit alarm system.

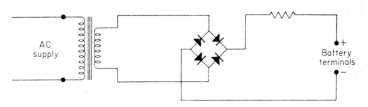

Figure 28.14. Circuit diagram of a constant-voltage battery charger.

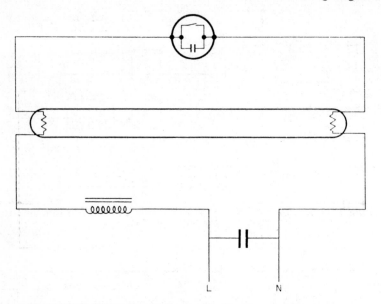

Figure 28.15. Switch-start fluorescent circuit with a glow-type starter.

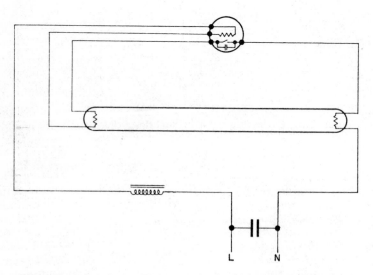

Figure 28.16. Switch-start fluorescent circuit with a thermal-type starter.

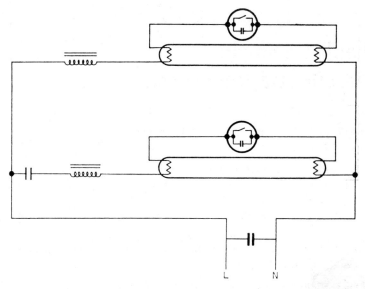

Figure 28.17. Wiring diagram for a lead-lag circuit for operating a pair of fluorescent lamps.

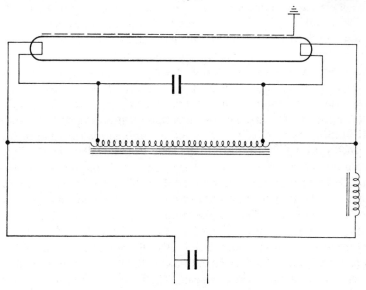

Figure 28.18. Instant-start circuit for a fluorescent lamp.

29

Fault-tracing in circuits and equipment

Types of fault
The types of fault which may occur in an electrical circuit fall into four general groups:

1. Open-circuit fault (loss of continuity).
2. Earth fault (low resistance between live conductor and earthed metalwork).
3. Short-circuit fault (low resistance between live and neutral conductors).
4. High-value series-resistance fault (bad joint or loose connection in conducting path).

These fault types occur in lighting and power circuits, appliances, apparatus and electric motors; variations do, of course, occur with the type of electrical equipment. Before any fault can be found and rectified it is necessary for the electrician to adopt a method or system based on a sound knowledge of circuitry and electrical theory, and on experience. The electrician detailed to repair a faulty circuit is in many ways like a doctor who makes his diagnosis on the basis of the symptoms revealed through a visual inspection or a test using the correct instruments. Haphazard tests carried out at random seldom lead to success in the quick location of faults. The investigation must always be based on an intelligent assessment of the fault and its probable causes, judged from its effects. In many instances, faults arise from installations or circuits which do not in some way or other comply with the requirements of the IEE Regulations, or else are used in such a manner that the abuse results in a fault. Most faults are easily located by following up reports such as 'There was a flash at the lamp'; 'The wires got red hot'; 'The lamp goes dim when it is switched on' or 'The bedroom light will come on only when the bathroom switch is ON'. By careful questioning, these reports will enable the electrician to locate the fault quickly and restore the circuit to normal operation again. The following are some common installation defects and omissions which eventually lead to faults:

1. The provision of double-pole fusing on two-wire systems with one pole permanently earthed. This frequently occurs with final sub-circuit distribution boards when the main and/or submain control equipment is single-pole and solid neutral.

2. Fuse protection not related to the current rating of cables to be protected. This is very often due to the equipment manufacturers fitting the fuse-carriers with a fuse-element of maximum rating for the fuse-units in the equipment.

3. Connecting boxes for sheathed-wiring systems placed in inaccessible positions in roof voids and beneath floors. Indiscriminate bunching of too many cables using screw-on or inadequate connections.

4. Insufficient protection provided for sheathed wiring, e.g. to switch positions and on joints in roof voids.

5. Incorrect use of materials, not resistant against corrosion, in damp situations (e.g. enamelled conduit and accessories and plain-steel fixing screws).

6. Inadequate or complete omission of segregation between cables and/or connections, housed within a common enclosure, supplying systems for extra-low voltage; or telecommunication and power and/or lighting operating at a voltage in excess of extra-low voltage.

7. Insufficient attention given to cleaning ends of conduit and/or providing bushings. Omission of bushings to prevent abrasion of cables at tapped entries, particularly at switch positions.

8. Insufficient precautions taken against the entry of water to duct and/or trunking systems, particularly where installed within the floor.

9. Incorrect use of TRS and PVC insulated and/or sheathed flexible cords instead of heat-resistant type, for connections to immersion heaters, thermal-storage block heaters, etc.

10. Incorrect use of braided and twisted flexible cords for bathroom pendant fittings and similar situations subject to damp or condensation.

11. The incorrect use of accessories, apparatus or appliances inappropriate for the operating conditions of the situation in which they are required to function. This often applies to agricultural and farm installations.

12. Installation of cables of insufficient capacity to carry the starting current of motors.

13. Incorrect rating of fuse-element to give protection to the cables supplying the motor.

Open-circuit faults

The instrument used to locate this type of fault is the continuity tester. The usual effect of this fault is that the apparatus or lamp in the circuit will not operate. The fault can be (*a*) a break in a wire; (*b*) a very loose or disconnected terminal or joint connection; (*c*) a blown fuse; (*d*) a faulty switch-blade contact. The fuse should always be investigated first. The rewireable type can be easily inspected. The cartridge type must be tested for continuity of the fuse-element. If the fuse has operated, the reason why it has done so must be found out. It is not enough to repair or replace the fuse and leave it at that. A broken wire or a disconnection will show on the continuity tester as an extremely high resistance in the kilohm or Megohm ranges. Before each wire in the faulty circuit is tested in turn (live feed, switch-wire and neutral) all mechanical connections should be inspected (lampholders, junction box, plug, switch and appliance terminals). Conduit, trunking or the metal sheathing of certain wiring systems can be used as a convenient return when testing the continuity of very long conductors. In an all-insulated wiring system, other healthy conductors can be used as returns for testing purposes, making sure that the original connections are restored once the fault has been found.

Earth faults

An earth fault between a live conductor and earthed metalwork will have the same effect as a direct short-circuit: the circuit fuse will blow. To trace the fault, it is necessary to isolate the live conductor from the neutral by removing all lamps, etc., and placing all switches in the ON position. An insulation-resistance (*IR*) tester is used to trace this fault. Circuits should be subdivided as far as is possible to finally locate the position of the fault. The reading obtained on the instrument used will be in the low-ohms range. An earth fault on the neutral conductor seldom shows up except by an *IR*-to-earth test on the neutral conductor. In most instances this type of fault does not affect the operation of the circuit or the devices or equipment connected to it. However, it is important to rectify any such fault found, otherwise if it is ignored it may cause a shock and fire hazard.

Short-circuit fault

On testing the insulation resistance between the live and neutral conductors with an *IR* tester, the reading will show itself in the low-ohms range. Again, subdivision of the installation at the distribution board, and subdivision of the faulty circuit, is the only way to locate and confirm the position of the faults. Short circuits can occur as the result of damaged insulation, bare wire in junction boxes and fittings, or by a conductor becoming loose from terminals and moving so as to come into contact with a conductor of opposite polarity. The result of a short circuit is a blown fuse, though if there is a sufficiently high resistance in the circuit (that is, not sufficient current can flow to blow the circuit fuse) the result will be overheating of the conductors and sparking or arcing at the point of contact. The test involves the removal of all lamps and appliances from the faulty circuit, open all switches, and carry out an *IR* test between the live and neutral conductors. If the reading obtained is satisfactory, close each circuit switch in turn until the faulty conductor, a switch wire, is located. If a low or near-zero reading is obtained on the first test, the circuit will have to be disconnected at convenient points until the faulty wire is isolated.

High-value series-resistance faults

This type of fault is most difficult to trace as it usually means that a connection, joint or termination has become loose. The effect of this is invariably 'dim lights' or a motor 'going very slowly and heating up'. In new installations the dimness of the lamps may well be caused by a wrong connection in a junction box resulting in two or more lamps being connected in series.

Main faults in new wiring

Faults in new wiring are generally the result of careless or inadvertent wrong connections which will either blow a fuse, cause lamps to operate dimly as above, not work at all or work only when another circuit switch is placed on the ON position. If a lamp lights only when another switch in the same final subcircuit is ON, this indicates that the live feed to the 'faulty' lamp has been looped from the switch-wire side of the previous circuit switch instead of from the live-feed side. The fact that overloading a circuit will blow a fuse should not be overlooked.

Faults in fluorescent-lamp circuits

The following tables summarise the faults, effects and the remedies associated with fluorescent-lamp circuits.

Table 29.1 *Fault-finding in fluorescent lamp fittings*

If a fitting fails to operate correctly, check as follows:

Step No.	Item	Tests to be applied
1.	Supply and fuse	Check supply voltage at input to fitting. Check polarity of incoming supply and ensure frame is earthed. If fuse has blown, suspect circuit or component and find the fault before replacing fuse.
2.	Lamp	Check lamp in a good fitting and if proved faulty replace with a new lamp. Remember, never try a new lamp in a fitting which has faulty components or circuit.
3.	Circuit	Examine wiring inside the fitting and if possible check against the wiring diagram. Check insulation resistance between the circuit and the metal frame of the fitting. The resistance should be above 2 megohms. If an earth fault is found, trace the cause and replace the component.
4.	Ballast chokes	Examine for signs of overheating, if possible check continuity of windings and insulation resistance. Compare the impedance or inductance against a good replica.
5.	Capacitors	Examine for leakage or damage. If possible check the capacitance and check that the discharge resistor has a value between $\frac{1}{4}$–1 megohm. The insulation resistance between case and terminals should be above 2 megohms.
6.	Starter switches	Check operation of starter in another good circuit and, if found faulty, fit a new replacement.
7.	Ambient conditions	Remember that normal fluorescent fittings may overheat if the surrounding temperature is above 30–35°C. Lamp starting may be difficult with some types of circuit if the temperature is below 5°C.

Table 29.2 *Quick-start circuits*

Symptom	Possible fault	Test and remedy
Lamp fails to start— both ends glowing brightly	Wrong type of lamp or inefficient earthing	Ensure that correct grade of lamp is fitted. Quick-start lamps must have earthed metalwork within 12mm of the tube along its full length.
	Low voltage	If supply voltage is below 220V make sure that the leads from the choke and neutral are connected to inner terminals of the quick-start transformer unit.
	High or low temperature	Ventilate fitting if excessive temperature. Screen or enclose fitting if low temperature, or use correct grade lamp.
Lamp fails to start—one end glows brightly	Broken cathode	Test lamp in sound fitting or special test transformer unit.
	Lampholder not making contact or short-circuited	Check lampholder contacts and leads for open or short circuit. Test output voltage from quick-start secondary.
Lamp fails to start—no end glow	Open circuit or short circuit on quick-start	Test voltage applied to fittings—if correct, check circuit for open-circuited choke or quick-start or for short circuit on quick-start.
Lamp fails to start— ends glow dull and reddish	Faulty lamp	Test lamp in sound fitting or in special test transformer.
	Low cathode heating	Check output voltage from quick-start secondary.

Note: Do not make assumptions; check first the simple things, i.e. fuses and switches. Do not leave voltmeters in circuit when closing switches.

Table 29.3 *Switch-start circuits*

Symptom	Possible fault	Test and remedy
Lamp does not attempt to start—no end glow	Fuse blown	Test voltage applied to fitting—trace broken fuse. A standard Avometer may be used to obtain voltage reading. Should voltmeters be used, it may be necessary to employ a 0–400V range for general testing, and a 0–20 or 0–25V range for cathode heating tests. In leading power factor circuits, the voltage across the capacitor will be approximately 400V. A test lead, with incandescent lamp of rated voltage, may be used to test for mains voltage at a fitting.
	Faulty starter	Insert starter switch in sound fitting or special test lead. A glow-switch type starter may be connected in series with a 25W 230V tungsten filament lamp across the main supply. The starter switch will then operate and cause the lamp to switch on and off.
	Faulty lamp	Insert tube in sound fitting or test each cathode in turn in special test transformer unit. A 12V quick-start transformer is useful for testing cathode emission. If the 12V winding is connected across a tube cathode, the lamp end will light up its normal colour indicating satisfactory operation. If cathode glows dull red then local ionisation is absent and indicates life-expired lamp. If one or both cathodes are broken, check for faulty circuit (short-circuit to earth or wrong control gear) before inserting new tube.
	Open circuit	Test for open circuit on choke, etc., or short to earth between choke and tube.

Table 29.3—*cont.*

Symptom	Possible fault	Test and remedy
Lamp fails to start— bright glow from one end	Crossed leads in twin lamp fitting	Check that the correct lampholders are connected to each tube, i.e. one or both cores connected to a given tube should have the same colour.
	Short circuit on lampholder lead	Test for short circuit across lampholder lead or for short circuit to earth on starter switch or wiring.
	Faulty lamp	Test for internal short circuits on cathode. A pen torch type of lamp tester can be used for cathode continuity reading. Alternatively a 0–30 ohm continuity tester can be used to check cathode resistance: 2–10 ohms (cold).
Lamp does not attempt to start—both ends glow brightly	Short circuit on starter switch or associated wiring	Test starter switch in sound fitting or special test lead. If satisfactory test starter switch socket and wiring for short circuit.
Lamp flashes on and off —fails to maintain discharge	Faulty lamp	Test lamp in sound fitting or special test transformer. At end of life other symptoms are reduced light output, increased flicker and reddish glow from cathodes.
	Low voltage or incorrect tapping	Test voltage at terminal block of fitting—if low, check external wiring for excessive voltage drop (fuse holders, etc.). Check choke tapping against 'sustained' voltage at terminals. If voltage is persistently low reduce tapping or convert to series capacitor circuit.
	Faulty starter	Test starter switch in sound fitting or special test lead.
	Low temperature	Screen open type fittings or use low-temperature grade lamps.
	Crossed leads in twin fitting	Check that the correct lampholders are connected to each lamp, i.e. one or both cores connected to a given tube should have the same colour.

Faults in motors and circuits

The following table summarises briefly the faults, effects and remedies associated with motors and their associated circuitry and control gear.

Table 29.4

Fault	Possible cause	Corrective action
1. Vibration	Uneven foundations	Check level and alignment of base and realign.
	Defective rotor	See (8).
	Unbalance	Uncouple from driven machine, remove motor pulley or coupling. Run motor between each of these operations to determine whether unbalance is in the driven machine, pulley, or rotor. Rebalance.
2. Frame heating*	Excessive load	See (5).
	Foreign matter in airgap or cooling circuit	Check airgap, dismantle motor and clean.
	Excessive ambient temperature	Motors supplied to BS2613 : 1957 are intended for operation in an ambient not exceeding 40°C. Where the ambient exceeds 40°C a motor of corresponding lower temperature rise should be used.
	Partial short or open circuit in windings	Check windings with suitable meter. If defective, repair or return to manufacturer.
3. Bearing heating	Too much grease	Remove surplus grease.
	Too little grease	Wash bearings and replenish with grease.
	Incorrect assembly	Ensure bearing assembled squarely on shaft.
	Bearing overloaded	This may be due to misalignment of the drive, excessive end thrust imposed on motor, or too great a belt tension. Take appropriate steps to reduce the load on the bearing.
4. Brushes heating	Excessive load	See (5).
	Brushes not bedding or sticking in holders	Carefully rebed or clean brushes and adjust pressure.
	Brush chatter	Ensure that commutator is true without high or low bars and adjust brush pressure.
	Incorrect grade of brushes	Ensure that brushes used are those specified by the motor manufacturer.

Table 29.4—*cont.*

Fault	Possible cause	Corrective action
5. No rotation	Supply failure, either complete or single phase	Disconnect motor immediately—with a single-phase fault serious overloading and burn-out may rapidly occur. Ensure that correct supply is restored to motor terminals.
	Insufficient torque	Check starting torque required and compare with motor rating, taking into account type of starter in use. Change to larger motor or to different type of starter.
	Reversed phase	Check connections in turn and correct.
6. Steady electrical hum	Running single phase	Check that all supply lines are alive with balanced voltage.
	Excessive load	Compare line current with that given on motor nameplate. Reduce load or change to larger motor.
	Reversed phase	Check connections in turn and correct.
	Uneven airgap	Check with feelers. If due to worn bearings fit new ones.
7. Pulsating electrical hum	Defective rotor	Check speed at full load. If it is low and if there is a periodic swing of current when running, a defective rotor is indicated, and the matter should be referred to the manufacturers.
	Defective wound rotor. Loose connection, partial short circuit, etc.	On a wound rotor machine check should be made of rotor resistance and open-circuit voltage between slip-rings.
8. Mechanical noise	Foreign matter in airgap	Check airgap, dismantle rotor and clean.
	Bearings damaged	Check with a listening stick. If confirmed try rotating outer race of bearing 180°. If still unsatisfactory fit new bearing.
	Couplings out of line	Check coupling gap and realign.

* The frame temperature should be checked with a thermometer; the reading so obtained will be approximately 10°C lower than the actual temperature of the windings.

Faults in low-voltage circuits

Because the voltage of low- and extra-low voltage circuits is relatively small, a poor or dirty contact will immediately prevent bells and similar devices from operating. These faults are thus most difficult to trace, and it is often a matter of systematic checking for continuity (zero or near-zero) readings. The prevention of faults on ELV circuits is more often than not a matter of regular periodic maintenance attention (cleaning contacts, tightening connections, etc.) than anything else.

Appliances and apparatus

The consumer-user range of appliances (cleaners, irons, kettles, fires and so on) have generally to withstand much handling which eventually leads to faults. One of the most common is a break in the continuity of a conductor of three-core flexible cord, particularly where the cord emerges from the plug or enters the appliance connector, or the appliance cable-entry position. This break in the conductor is most often the result of excessive movement of the cord at one point over a long time. The fault is usually identified by the intermittent working of the appliance and perhaps signs or smell, of burning. Conductors are often pulled away from their terminals to cause short circuits or intermittent high series-resistance contacts.

30

Estimating materials

Before any electrical work is attempted, it is necessary to find out (a) how much the materials are going to cost and (b) how long the job will take. Thus, an estimate is needed so that a total figure can be given to a customer who then decides whether the price is right for him, when compared with the prices submitted by other electrical contractors. An estimate contains three elements: materials, labour and overheads.

Materials bought by an electrical contractor are generally bought in bulk at wholesale or trade prices, and then sold to the client at a competitive retail price. The materials must, of course, be of good quality; it is part of the job of the estimating engineer to see that the best materials are used, the cost of which will enable a profit to be made when sold to complete an installation.

Labour: every job must be performed by workmen of various skills and abilities. The more skilled the craftsman, the more he must be paid. Some classes of work can be done by semi-skilled workers; other classes require the attention of an advanced craftsman or technician, if not actually to do the job, then to supervise others. Thus, the labour cost is an important element in an estimated price for a contract.

Overheads: any electrical contracting firm of reasonable size requires the services of people who specialise in clerical work, typing, storekeeping, drafting, accountancy and so on, who, though they are not directly associated with actual contracting work, still play an important part in the efficient operation of the firm. Also, premises must be provided for offices, showrooms, stores and workshops; and these must be heated and lit. Special tools, instruments, office equipment and transport are also needed. All these aspects are classed as overheads and the profit made on a job must help to pay for them.

To find out the amount of materials required to do a job it is necessary first to inspect the job site and measure the most probable runs of cable, the types of wiring accessories and the fittings required. To do this estimating work accuracy is required which can only come with some years of experience. The cost of the materials is obtained from the catalogues of manufacturers, which usually quote a wholesale or trade price and the recommended retail price. Estimating in its simplest form takes account of the amount of materials needed, the cost of these materials, and the time taken to do the job. Comparison of wiring methods on housing sites:

Wiring system	Cost (per cent)	Man and mate hours on site
Screwed conduit	157	634
Grip continuity conduit	153	500
MICC	150	220
TRS	100	167
Harness with metallic braiding	105	100

The time taken to do a particular job is again a matter of experience: knowing just how long an electrician will take, for instance, to raggle or channel a wall to contain a conduit switch-drop. If there are twenty such drops and each channel takes thirty minutes then the total time is $20 \times 30 = 600$ minutes, or 10 hours at so much per hour.

So far as the apprentice electrician is concerned, he may from time to time be asked to prepare a requisition for wiring materials to be used on small jobs, usually appliance repairs, or minor additions or alterations to an existing installation. The following are two general examples of schedules of materials, with no account taken of their cost (obtained from makers' catalogues).

Example 1

A final subcircuit is to be taken from a spare way in a distribution fuseboard. The circuit comprises 6 lighting points with 6 switches. The average run per point is 7 m. Each point terminates in a ceiling rose with flexible pendants. The cable to be used is PVC-insulated and sheathed. Fixings are surface on wood throughout.

Cable. 1 mm² flat PVC Twin & Earth, 250V grade—6 points×7 m plus 2 per cent for cable wastage $=42 + 1 = 43$ m (in practice a 50 m drum of cable would be taken to the job).

Flexible cord for pendants. 6 × 0·3 m = 1·8 m or 2 m of 0·5 mm² twin-core twisted flexible cord, VR-insulated, rayon-braided; colour maroon.

Switches. Six screw-on cover tumbler switches, one-way, 5A rating, moulded, for surface mounting; colour brown.

Ceiling roses. 6 × 2-plate moulded ceiling roses, for surface mounting; colour brown.

Pattress. 12– to suit switches and ceiling roses.

Lampholders. 6 × 2-slot bayonet cap moulded lampholders with metal liners; cord-grip type with shade carrier ring; colour brown. (Note that shielded lampholders may be used.)

Fixing clips. See Table B2M in the IEE Regulations which indicates the spacings of clips for horizontal runs (250 mm) and vertical runs (400 mm). Assuming a height to the ceiling of 2·75 m and the switches are placed at 1·5 m from finished floor level then 6 switches at 1·25 m = 7·5 m.

Therefore

$$\text{the number of clips (vertical runs) } \frac{7·5 \times 1000}{400} = 19$$

Remainder of run is largely on the horizontal: 43 m − 7·5 m = 35·5 m
Therefore

$$\text{number of clips (horizontal runs) } \frac{35·5 \times 1000}{250} = 142$$

Therefore, total number of buckle clips to contain 1 mm² flat PVC Twin & Earth = 142 + 19 = 161, with 12 mm in brassed iron fixing pins. (In practice, two 100-packs would be taken to the job.)

There must also be sundry fixings such as wood screws for switches, wood blocks, and ceiling roses. Lamps may have to be supplied. Joint boxes (insulated) will be required. Otherwise the feed for each switch would have to be looped. Allow for four joint boxes on this job.

Example 2

The following is a list of materials required for installing a circuit to a three-phase, 3 kW motor controlled by a star-delta starter. The cable is to be 4·00 mm² PVC, carried in conduit. The motor and starter are already on site. The length of run to the starter is 105 m (assume 6 right-angle bends in the run). Length of run from starter to motor is 3 m.

Material	Quantity
Main switch and fusegear	
30A triple-pole switchfuse	1
Starter is already on site	1
25 mm black-enamelled solid elbows	2
25 mm internal brass bushes	2
Cable and conduit included in the following section	
Sundries: Rawlplugs, screws, etc.	
Fuseboard to isolator	
20 mm black-enamelled screwed and welded conduit	117 m
20 mm black-enamelled couplers	6
20 mm black-enamelled sockets	12
20 mm black-enamelled spacer-bar saddles	60
20 mm internal brass bushes	4
20 mm black-enamelled locknuts	12
4·00 mm^2 red PVC cable	110 m
4·00 mm^2 yellow PVC cable	110 m
4·00 mm^2 blue PVC cable	110 m
Sundries: Rawlplugs, screws, etc.	
Starter to motor	
25 mm black-enamelled screwed and welded conduit	3 m
25 mm flexible conduit	50 mm
25 mm flexible adaptors	2
25 mm black-enamelled conduit box	1
25 mm reducer	1
25 mm black-enamelled sockets	3
25 mm black-enamelled locknuts	4
2·5 mm^2 red PVC cable	8 m
2·5 mm^2 yellow PVC cable	8 m
2·5 mm^2 blue PVC cable	8 m
2·5 mm^2 earth wire	1 m
Sundries: Rawlplugs, screws, etc.	

31

Report writing

Language is used for two purposes. First, it is used to communicate our ideas in speech or in writing, so that other people can know what we are thinking. Secondly, it enables us to receive from other people their own ideas. The result is that progress in many fields of human activity can be made. For instance, you receive an instruction to do a small wiring job. You acknowledge that instruction, either in writing or verbally. You communicate to the storekeeper your material requirements. Because he understands your communication (your Materials Requisition form) he gives you the right materials. With these materials you complete the job, and tell your supervisor. Thus, through various communication media something has been done. In order that complete understanding can be a feature of communications, it is important that what we say or write is clear; and we must examine carefully what other people have said or written. It has been estimated that a man working on the land, with a restricted education, has a vocabulary of less than a thousand different words. The average man, with a reasonable education, uses up to three thousand different words. This means that the latter man is more able to make himself understood to his fellows than the land worker. Thus, it is important to know enough words—which can come with the extensive reading of almost anything.

So far as the technical student is concerned, he finds himself in a world which uses an extension of normal language. This extension is usually called 'technical jargon'. Each trade has its own words, and the electrical engineering field is no exception. There are many thousands of special words which mean nothing to the layman, but to the knowledgable electrical worker they mean something. And when they are used in conversation between two or more electrical people, a basis for common understanding is established. Language to the technical student is just as much a 'tool' as is a screwdriver or a pair of pliers. It performs a definite function. Thus, the best language to use in technical fields (descriptions of electrical apparatus, technical reports and the like) is functional English, that is English which performs the function of establishing a definite understanding between two or more people. Before one can use words, one must know their correct meaning.

Writing technical descriptions, of electrical apparatus, wiring accessories, cables and other items, requires careful observation of the subject. Technical description is more difficult because it is more important to be clear and exact in this type of writing than in other types. When describing a piece of apparatus one must know what one is talking about: what it is made from, its function, its characteristics and how it may differ from pieces of apparatus with similar functions, say, cartridge and rewireable fuses. An observing eye is needed, as well as an ability to select those aspects which will give a reader, even a lay reader, a full picture of the subject.

Because many examination questions require written answers, it is important that every technical student makes himself familiar with language and particularly those words associated with his career subject. Not everyone, of course, can express himself clearly. But despite this, a full description can be given of an item without a great deal of what is called 'literary ability'.

Every technical description must have a plan, as follows. First, the introduction which gives the name of the object, the purpose for which it is used, and the principle on which it operates. Secondly, there is a description of the object. It is useful in this part to divide and subdivide the object into its various main and subparts. The amount of detail will depend on the required length of your description, and the complexity of the object. Everything essential should be included in the description. Indications should be made of the purpose of the most important parts, the material from which they are made, and their size, shape, weight, strength and so on. Thirdly, there should be a description of how the object works. This is very important when describing machinery.

For those whose command of language is not very good, the use of illustrations can help immensely. The Chinese have been credited with the saying that 'A good picture is worth a thousand words'. And so it is. For should the reader also not have a good command of language, the illustration will help him to understand what the object is, what it looks like and often its purpose. An illustration also saves time, not only in writing but in reading. Imagine how difficult it would be in practice if in a maker's catalogue, every switch had to be described in words instead of through the use of drawings drawn by a draughtsman or even a photograph. A point to remember in examination questions is that if the question asks definitely for a description, then the words are more important than any illustration used to supplement the description.

The main points to keep in mind about a technical illustration are:

1. Keep it simple.
2. Keep it tidy.
3. Make sure it is large enough to be legible.
4. Label all the parts.
5. Title the drawing.

To save time in practice, much use is made of forms which ask all the relevant questions, and the writer need only answer the questions in a short concise manner. Examples of two Test Certificates are shown in the IEE Regulations: the Completion Certificate for an electrical installation, and an Inspection Report on the condition of an installation. These reports are simple to fill in. Where a certain amount of literary ability enters is in that part of the report which requires details of departures from standard practice, or defects and faults found in the installation. The recommendations by the writer as to the best way to remedy the faults or defects must be easily understood.

A typical report on the condition of an electric motor is given in Appendix A of this book.

Part B

Electrical workshop technology

32

Tools

There is virtually no job that can be done without the use of tools. Tools can be regarded as extensions to our arms and fingers, in much the same way as the computer is an extension to the human brain. Tools enable us to perform many tasks which would otherwise be extremely difficult, such as tightening up a nut with the fingers instead of with a spanner. No doubt the nut could be tightened up with the aid of a few bits and pieces, if one were inventive enough. But the spanner is designed to do the job so why use anything else? Again, how many screwdrivers are used as chisels? Pliers as hammers? And even hammers as screwdrivers? The good workman recognises his tools as items without which he cannot do a good job. So far as the practicing electrician and electrical craftsman is concerned, he is expected by his employer to build up, through his years of apprenticeship, the following kit of tools:

1 ratchet brace and set of bits
1 large and 1 small screwdriver
2 pairs of cutting pliers
1 1-metre folding rule
1 adjustable hack-saw frame
1 pair 180 mm 'Footprints' and 1 pair 230 mm 'Footprints'
1 bradawl
1 wood chisel
1 pad-saw
2 hammers
1 plumb-bob and line
1 knife
1 spirit-level
1 tenon-saw
1 centre punch
1 wheel-brace

1 pair sidecutters
1 set of spanners ⅛ in to ½ in Whitworth (or Metric equivalent)
1 set of box spanners ⅛ in to ¾ in Whitworth (or Metric equivalent)
1 adjustable tap wrench
1 miniature 150 mm hack-saw frame
2 cold chisels, small and large, or (where appropriate):
1 cold chisel and 1 tongueing chisel.

The electrician's adult assistant is required to supply himself with a knife, a bradawl, a small screwdriver and a pair of cutting pliers.

The electrical fitter, who works more in the workshop, will have a slightly different set of tools, though basically the tools are designed for performing definite electrical tasks. The following are some of the tools used in electrical work, and general workshop practice.

The chisel

The chisel is used in a number of ways in the workshop, but mainly to cut or to shape metal. Outside the workshop, it is used for cutting away concrete, brick and other building materials to make, way for cable and conduit runs. The action of the chisel is more a tearing than a cutting process. Cutting effectively with a chisel requires a knowledge of the material to be cut; that is, the angle or rake and clearance must be considered. Cutting hard metals calls for less rake than for soft metals; the chisel angle must be less for softer metals. The following table gives suitable angles:

Cast-steel	65°
Cast-iron	60°
Mild-steel	55°
Brass	50°
Copper	45°
Aluminium	30°

The flat chisel is the general purpose tool. It is tapered and flattened for about one-third of its length to the cutting edge, which should be about 3 mm thick on the large chisel and less in proportion for smaller chisels. The cutting edge is ground to an angle suited to the material to be worked; it should not be exactly straight, but given a slight curve. On no account should the head of a chisel be hard where it is struck by the hammer. The cross-cut chisel is for grooving material, cutting keyway slots, and for entering places where the flat chisel is not a convenient tool.

The hack-saw

This is the main tool used for cutting-off and other workshop operations. There are many types and sizes of hack-saw. Blades may be of carbon or high-speed steel, and may be finished either with only the cutting edge hardened, or they may be hardened right through. The most usual blade for hand work is about 250 mm long by 12 mm wide. Soft-backed blades are tougher and are less liable to snap than the all-hard blades, though they are not so efficient cutters. Hack-saw teeth are specified by the number of teeth per inch, the pitch of the teeth being the reciprocal of this number (e.g. blade with 16 teeth per 2·5 cm has a pitch of 1·5 mm). The teeth on a hacksaw blade have a set to the sides. This causes the blade to cut a wider slit than itself, and prevents the body of the blade from rubbing or jamming in the saw-cut. Fine-toothed blades, for cutting thin metal, are sometimes made with a wavy set to minimise stripping of the teeth from the blade. The best all-round blade for hand use is the 16–18 teeth per 2·5 cm. For other and special classes of work the following blades should be used:

solid brass, copper and cast-iron: 14 teeth per 2·5 cm

silver steel and thin cast-steel rods; thin structural sections: 22 or 24 teeth per 2·5 cm.

sheet metal and tubing (e.g. steel, copper and conduit tubing): 32 teeth per 2·5 cm.

The file

This is a cutting tool. The most usual form of file is that with cross-cut teeth; that is, the grooves in the face of the file run in two directions dividing it up into small diamond-shaped teeth. The single-cut file has only one series of grooves. There are many types of file:

The *flat file* is parallel for about two-thirds of its length and then tapers in width and thickness. It is cut on both faces and both edges. The *hand file* has parallel width throughout its length; its thickness tapers like that of the flat file. Both faces are double cut. One edge only is single cut. The 'safe' edge (the uncut edge) prevents cutting into one face of a square corner while the other face is being filed. Both the hand file and the flat file are used for general surfacing work. The hand file is used particularly when filing up to a step which must be kept straight and square. The *square file* is parallel for two-thirds of its length and then tapers off. It is double-cut on all sides and is used for filing corners and slots, where a hand file cannot enter.

The *round file* tapers as does the square file. It is used for opening out holes, producing rounded corners, round-ended slots and so on. The rounded side of the *half-round file* is not a true half-circle. This side of the file is useful for producing a radius. The *three-square file* is used for corners less than 90°, and in positions where awkward corners have to be taken out. Files are graded as rough; bastard; second cut; smooth; and dead smooth.

The scraper

The flat scraper resembles the file but has no teeth. Metal is removed from a surface by scraping rather than cutting.

The vice

The vice (or vise) is used for holding work while it is being operated on. There are many types of vice. For the electrical workshop there is the bench vice and the machine vice, the latter being used to hold items while they are being drilled. The bench vice is used for benchwork. The pipe vice is used for holding pipework and is probably the more familiar vice to the electrician outside the workshop.

The hand hammer

The most used of all tools is the hammer. There are many shapes of hammer heads, each designed for a specific type of work. They are usually classed by the shape of the end opposite the striking face, called the pein. The ball pein is common; the ball end being used mostly for riveting over the ends of pins and rivets. Cross and straight peins are also useful for riveting in awkward places. The main consideration in choosing a hammer is that of its weight. Engineers' hammers are made in weights varying from 100 g to $\frac{1}{2}$ kg. The most common weights in the range are from 250 g to 750 g or 1 kg. Light hammers are used for lighter and more delicate operations such as light riveting, striking centre punches, and small chisels, driving in small pins, etc. Heavy hammers are necessary to do heavy chipping, driving pulleys and collars onto shafts and driving large pins into holes. Hammer shafts should be made from well-seasoned, straight-grained hickory or ash, and be free from knots and other defects. The length of the shaft should be suited to the size of the head.

The centre punch

This tool is used when circular dot marks are required. When a job has to be marked out it is usual to follow along the lines with small dot marks in case the lines are rubbed away. The surface to be finished is taken to the centre of the dot mark. Centre punches are also used for marking the centre point of drilled holes to give the drill a good start, and for marking the centre of circles to provide a suitable point for placing one leg of dividers to scribe the circle.

Use and care of tools

All tools should receive regular care and attention at frequent intervals throughout their working life. This is necessary if the tools are expected to perform their tasks efficiently. As a tool edge is used it becomes dull and needs sharpening. Hammer-heads become loose and require tightening. Screwdriver blades become distorted and tapered and need to be reground carefully to restore an efficient working edge. Twist drills require grinding correctly for quick, clean and accurate work. Tools used as gauges should be kept aside when not in use so that their surfaces are not damaged in any way.

When soft metals are filed, the file teeth tend to become clogged with minute lumps of metal. If the teeth are not cleared of this, the filing operation will prove difficult. If the metal is not too firmly wedged, it may be removed using a file card, which is a brush made from a strip of webbing with thin, hard-wire bristles nailed onto a piece of wood. Tightly-wedged lumps of metal must be laboriously picked out with the point of a scriber. File teeth are very brittle. Thus, files should never be heaped together, but kept separate.

33

Workshop safety

Safety in an electrical workshop means much more than the provision which enables persons to use electricity without risk of shock. Means to prevent accidents, not necessarily of an electrical nature, must at all times be part and parcel of workshop procedure and facilities. There are two main causes of accidents in factories and workshops. First, carelessness. When any task is done often enough it becomes familiar to the point at which boredom sets in; when this happens, original high standards begin to slide. The result is often an accident to that person or to an innocent fellow-worker, which may or may not be fatal. Secondly, there is the ignorance factor. This does not mean lack of intelligence. Rather it means a lack of sufficient technical knowledge to perform a task efficiently and to leave the task in a condition which is safe. Many tasks seem simple when done by the expert. If there is a lack of knowledge, a task should be done under expert supervision.

Electric shock being one of the main occupational hazards of the electrician, the following points should be observed:

1. Never take it for granted that a circuit is 'dead'. Always check thoroughly using a test lamp or a neon tester. Never check for 'live' by touching conductors with fingertips.
2. If a circuit must be worked on while 'live', it is always best to leave the work to a more experienced electrician. Otherwise, use adequately-insulated tools, rubber gloves, and stand on rubber matting. The higher the voltage the greater the care needed.
3. If working on a circuit to which other persons may have access, always remove the circuit fuses and indicate that the circuit has been made 'dead' deliberately. Keep the fuses in your pocket until the job is finished.

So far as general care in the workshop is concerned, always be careful when working on or beside moving machinery. Goggles must be worn when grinding metal. All machinery should have guards. Ladders should be placed at reasonable angles to the vertical and be placed on a non-slip surface.

First aid

Though first aid is meant only to be a temporary measure until skilled assistance can be obtained, it is no less important to have some basic knowledge of the treatment of injuries. Every factory and workshop must have a first-aid box; this is a legal requirement.

Treatment for electric shock

There are two methods used to treat electric shock. In any case, speed is essential as even a few minutes' delay may cause death. A severe electric shock will affect the nerves which control the breathing and the action of the heart. In the treatment, effort is made to get the patient's body working normally, even while unconscious.

Before the treatment is begun, it is important to make sure that the patient is not still in contact with the live object. If this is the case, then the patient must not be moved with bare hands, but should be pulled away from the live contact using a jacket, a chair, dry folded newspaper, a belt, rope or a length of dry wood. Standing on newspapers, dry wood, or dry clothes also helps to increase the insulation factor between the rescuer and the floor which may be conducting. Any obstructions to the patient's breathing should be removed (e.g. tight collar). Artificial respiration should be given immediately the patient is clear from the live electrical contact.

Holger Nielsen method. Place the patient face downwards. See that the forehead is resting on the hands (placed one over the other) so that the nose and mouth are clear. A rolled-up jacket will help to raise the head. If the patient's tongue has been swallowed, two or three firm slaps with the flat of the hand on the area between the shoulders will bring it forward.

The rescuer should now kneel in front of the patient, a knee beside the right of the head and a foot to the left at the patient's elbow. The arms should slope forward so that the hands lie close together on the patient's shoulders; the wrists should be over the top of the shoulder-blades. Begin the movements as follows:

1. Move forward and press down with a light pressure to drive air from the lungs. This movement should last about two seconds (count: ONE, TWO).
2. Slide hands quickly down to patient's elbows, this movement taking about a second (THREE).

3. Raise the elbows slightly. The rescuer's body should move backwards a little. This movement induces air into the patient's lungs and should take about two seconds (FOUR, FIVE).
4. Lower elbows and slide hands to patient's back to resume original position (SIX).

The above movements should be repeated until the patient recovers. The movement allows for about nine respirations per minute. It is recommended that efforts to save life should be maintained for about four hours. Patients who have recovered should always be seen by a doctor, who should be sent for as soon as the accident has been discovered.

Mouth-to-mouth method. This method is sometimes known as the 'kiss-of-life' method of shock-treatment, and has proved easy to apply and as effective as the Holger Nielsen method.

1. Place the patient on his back and sit or kneel by the side of his head. The head should be held in both hands. One hand should press upwards; the other pushing the jaw upwards and forwards. This position is to ensure that the patient's lungs get maximum air to them.
2. Close the patient's nose with one hand.
3. Seal your lips round the patient's mouth. Blow air steadily into the patient's mouth until the swelling lungs cause the chest to rise.
4. Remove your mouth. Turn your head aside and take a deep breath.

Repeat this cycle six times as quickly as possible. Then continue at about 10-second intervals.

As a general precaution, a thin handkerchief may be placed over the patient's mouth or nose if desired. If the patient's lungs seem to be obstructed, this could well be the tongue which may have slipped back. The patient's head should always be kept pressed back.

Electric burns

These are best seen by a first-aid expert or by a doctor. Burnt clothing should not be removed. And blisters should not be broken. Sterilised dressing may be applied if thought necessary. Patient may be given water, tea, coffee or some other liquid, but no alcohol. Keep the patient warm at all times.

34

Electrical materials 1: Conductors

In electrical work, materials used fall into five general groups: conductors, semi-conductors, insulators, magnetic and constructional. Each group finds its place in the equipment, circuitry and devices to be found in all aspects of the electrical engineering field. Some materials are found different purposes. Iron, for instance, is a magnetic material and is also used for constructional purposes: and with other elements to form a ferrous alloy, it is a resistive conductor as found in heating elements. Chapter 35 deals with insulating materials.

Conductors

In electrical work, a 'conductor' means a material which will allow the free passage of an electric current along it, and which presents negligible resistance to the current. If the conducting material has an extremely low resistance (e.g. a copper cable) there will, normally, be no effect when the conductor carries a current. If the conductor material has a significant resistance (e.g. iron wire) then the conductor will show the effects of an electric current passing through it, usually in the form of a rise in temperature to produce a heating effect. It should be remembered that the conduction of electric currents is offered not only by metals, but by liquids (e.g. water) and gases (e.g. neon). Conductors by nature differ so enormously from insulators in their degree of conduction that the materials which offer strong resistance to an electric current are classed as insulators. Those materials which fall in between the two are classed as semi-conductors (e.g. germanium).

Copper

This metal has been known to man since the beginning of recorded history. Copper was connected with the earliest electrical effects such as, for instance, that made by Galvani in 1786 when he noticed the curious behaviour of frogs' legs hung by means of a copper hook from an iron railing (note here the two dissimilar metals). Gradually copper became known as an electrical material; its low resistance established it as a conductor. One of the first applications of copper as a conductor was for the purpose of signalling; afterwards the commercial generation of electricity looked to copper for electrical distribution. It has thus a prominent place and indeed is the first metal to come to mind when an electrical material is mentioned. As a point of interest, the stranded cable as we know it today has an ancient forebear. Among several examples, a bronze cable was found in Pompeii (destroyed AD 79); it consisted of three cables, each composed of fifteen bronze wires twisted round each other.

Copper is a tough, slow-tarnishing and easily-worked metal. Its high electrical conductivity marks it out for an almost exclusive use for wires and cables, contacts, and terminations. Copper for electrical purposes has a high degree of purity, at least 99·9 per cent. This degree of purity results in a conductivity value only slightly less than that of silver (106 to 100). As with all other pure metals, the electrical resistance of copper varies with temperature. Thus, when there is a rise in temperature, the resistance also increases. Copper is available as wire, bar, rod, tube, strip and plate. Copper is a soft metal; to strengthen it certain elements are added. For overhead lines, for instance, copper is required to have a high-tensile strength and is thus mixed with cadmium. Copper is also reinforced by making it surround a steel core, either solid or stranded.

Copper is the basis of many of the cuprous alloys found in electrical work. Bronze is an alloy of copper and tin. It is fairly hard and can be machined easily. When the bronze contains phosphorous, it is known as phosphor-bronze, which is used for spiral springs. Gunmetal (copper, tin and zinc) is used for terminals. Copper and zinc become brass which is familiar as terminals, cable legs, screws and so on, where good conductivity is required coupled with resistance to wear. Copper oxidises slowly at ordinary temperatures, but rapidly at high temperatures; the oxide skin is not closely adherent and can be removed easily.

Aluminium

The use of aluminium in the electrical industry dates back to about the turn of this century when it was used for overhead-line conductors. But because in the early days no precautions were taken to prevent the corrosion which occurs with bimetallic junctions (e.g. copper cable to aluminium busbar) much trouble was experienced which discouraged the use of the metal. Generally speaking, aluminium and its alloys are used today for electrical purposes because of (*a*) weight; (*b*) resistance to corrosion; (*c*) economics (cheaper than copper); (*d*) ease of fabrication;

(*e*) non-magnetic properties. Electrical applications include cable conductors, busbars, castings in switchgear, and cladding for switches. The conductor bars used in the rotor of squirrel-cage-induction ac motors are also of aluminium on account of the reduced weight afforded by the metal. Cable sheaths are available in aluminium. When used as conductors, the metal is either solid or stranded.

An oxide film is formed on the metal when exposed to the oxygen in the atmosphere. This film takes on the characteristics of an insulator, and is hard enough to withstand some considerable abrasion. The film also increases the corrosion-resisting properties of aluminium. Because of this film it is important to ensure that all electrical contacts made with the metal are initially free from it; if it does form on surfaces to be mated, the film must be removed or broken before a good electrical contact can be made in a joint. Because the resistivity of aluminium is greater than that of copper, the cross-sectional area of the conductor for a given current-carrying capacity must be greater than that for a copper conductor.

Zinc

This metal is used mainly as a protective coating for steel and may be applied to the steel by either galvanising, sherardising or spraying. In electrical work it is found on switchgear components, conduit and fittings, resistance grids, channels, lighting fittings and wall brackets. Galvanising is done by dipping iron or steel objects into molten metal after fluxing. Mixed with copper, the zinc forms the alloy brass. Sherardising is done by heating the steel or iron object to a certain temperature in zinc dust, to result in an amalgamation of the two metals, to form a zinc-iron alloy.

Lead

Lead was one of the oldest metals known to man. Lead is highly resistant to corrosion. So far as the electrical application of lead is concerned, apart from its use in primary and secondary cells, cable sheathing in lead was suggested as early as 1830–45. This period saw the quantity production of electrical conductors for inland telegraphs, and thoughts turned to the possibility of prolonging the life of the conductors: the earliest suggestions were that this could be done by encasing them in lead. Today lead is used extensively. Lead is not used pure; it is alloyed with such metals as tin, cadmium, antimony and copper. Its disadvantage is that it is very heavy; it is also soft, even though it is used to give insulated cables a degree of protection from mechanical damage. One of its principal properties is its resistance to the corrosive effects of water and acids. It has a low melting point; this fact is made use of in the production of solder, where it is alloyed with tin for cable-jointing work. Lead alloyed with tin and copper is used as white metal for machine bearings.

Nickel

This metal is used in conjunction with iron and chromium to form what is known as the resistive conductors used as heating elements for domestic and industrial heating appliances and equipment. The alloy stands up well to the effects of oxidation. Used with chromium only the alloy is non-magnetic; with iron it is slightly magnetic. It has a high electrical resistivity and low temperature coefficient. The most common alloy names are Nichrome and Brightray and Pyromic. Pure nickel is found in wire and strip forms for wire leads in lamps, and woven resistance mats, where resistance to corrosion is essential.

Carbon

This material is used for as motor brushes (slip-ring and commutator), resistors in radio work and contacts for some types of circuit-breaker. It has a negative temperature characteristic in that its resistance decreases with an increase in temperature.

Ferrous metals

These metals are based on iron and used for the construction of many pieces of equipment found in the electrical field (switches, conduit, cable armouring, motor field-poles and so on). Because iron is a magnetic material, it is used where the magnetic effect of an electrical current is applied to perform some function (e.g. in an electric bell).

The choice of magnetic materials today is extremely wide. For practical purposes magnetic materials fall into two main classes: permanent (or hard) and temporary (or soft). Permanent magnetic materials include tungsten and chromium steel and cobalt steel: when magnetised they retain their magnetic properties for a long time. Cobalt-steel magnets are used for measuring instruments, telephone apparatus, small synchronous motors. Soft magnetic materials do not retain their magnetism for any appreciable time after the magnetising force has been withdrawn. In a laminated sheet form they are found in transformer cores and in machine poles and armatures and rotors. Silicon-iron is the most widely used material for cores.

Rare and precious metals

In general, precious metals are used either for thermocouples or contacts. Among the metals used are silver, gold, platinum, palladium and iridium. Sometimes they are used as pure metals, otherwise as an alloy within the above group or with iron and copper, where special characteristics are required. For instance, a silver-iron alloy contact has a good resistance to sticking and is used in circuits which are closed with a high inrush (e.g. magnetising currents associated with inductors, electromagnets and transformers). It is used also for small motor-starter contacts. The alloy maintains low contact resistance for very long periods. The following are some applications of rare and precious metals in contacts:

Circuit-breakers. Silver, silver-nickel, silver-tungsten.

Contactors. Silver, silver-tungsten.

Relays. Silver, platinum, silver-nickel.

Starters. Platinum, rhodium, silver, coin silver. Silver is used for the fuse-element in HRC fuses.

Mercury. This material is used almost exclusively for mercury switches (see Chapter 21), and in mercury-arc rectifiers. In a vapour form it is used in fluorescent lamps (low-pressure lamps) and in the high-pressure mercury-vapour lamp.

Semi-conductors

Oxides of nickel, copper, iron, zinc and magnesium have high values of resistivity; they are neither conductors nor insulators, and are called semi-conductors. Other examples are silicon and germanium. When treated in certain ways, these materials have the property of being able to pass a large current in one direction while restricting the flow of current to a negligible value in the other direction. The most important application for these materials is in the construction of rectifiers and transistors.

Conducting liquids

Among the liquids used to conduct electric currents are those used as electrolytes: sulphuric acid (lead-acid cells); sal ammoniac (Leclanché cells); copper sulphate (in simple cells); caustic potash (nickel-cadmium cells). When salts are introduced to water the liquid is used as a resistor.

Conducting gases

In electrical work, so far as the practical electrician is concerned, the conducting gases are those used for electric-discharge lamps: neon, mercury vapour, sodium vapour, helium; the latter is used in the 'glow' type starter used in fluorescent starting circuits.

35

Electrical materials 2: Insulators

An insulator is defined as a material which offers an extremely high resistance to the passage of an electric current. Were it not for this property of some materials we would not be able to apply electrical energy to so many uses today. Some materials are better insulators than others. The resistivity of all insulating materials decreases with an increase in temperature. Because of this, a limit in the rise in temperature is imposed in the applications of insulating materials, otherwise the insulation would break down to cause a short circuit or leakage current to earth. The materials used for insulation purposes in electrical work are extremely varied and are of a most diverse nature. Because no single insulating material can be used extensively, different materials are combined to give the required properties of mechanical strength, adaptability and reliability. Solids, liquids and gases are to be found used as insulation.

Insulating materials are grouped into classes:

Class A—Cotton, silk, paper and similar organic materials; impregnated or immersed in oil.
Class B—Mica, asbestos, and similar inorganic materials, generally found in a built-up form combined with cement binding cement. Also polyester enamel covering and glass-cloth and micanite.
Class C—Mica, porcelain, glass, quartz and similar materials.
Class H—Silicon-glass.
Class E—Polyvinyl acetal resin.

The following are some brief descriptions of some of the insulating materials more commonly found in electrical work.

Rubber

Used mainly for cable insulation. Cannot be used for high temperatures as it hardens. Generally used with sulphur (vulcanised rubber) and china clay (for TRS sheathing). Has high insulation-resistance value.

Polyvinylchloride (PVC)

This is a plastics material which will tend to flow when used in high temperatures. Has a lower insulation-resistance value than rubber. Used for cable insulation and sheathing against mechanical damage.

Paper

Must be used in an impregnated form (resin or oil). Found used for cable insulation. Impregnated with paraffin wax, paper is used for making capacitors. Different types are available: kraft, cotton, tissue, pressboard.

Glass

Used for insulators (overhead lines). In glass fibre form it is used for cable insulation where high temperatures are present, or where areas are designated 'hazardous'. Requires a suitable impregnation (with silicone varnish) to fill the spaces between the glass fibres.

Mica

This material is used between the segments of commutators of dc machines, and under slip-rings of ac machines. Used where high temperatures are involved such as the heating elements of electric irons. It is a mineral which is present in most granite-rock formations; generally produced in sheet and block form. Micanite is the name given to the large sheets built up from small mica splittings and can be found backed with paper, cotton fabric, silk or glass-cloth or varnishes. Forms include tubes and washers.

Ceramics

Used for overhead-line insulators and switchgear and transformer bushings as lead-ins for cables and conductors. Also found as switch-bases, and insulating beads for high-temperature insulation applications.

Bakelite

A very common synthetic material found in many aspects of electrical work (e.g. lampholders, junction boxes), and used as a construction material for enclosing switches to be used with all-insulated wiring systems.

Insulating oil

This is a mineral oil used in transformers, and in oil-filled circuit-breakers where the arc, drawn out when the contacts separate, is quenched by the oil. It is used to impregnate wood, paper and pressboard. This oil breaks down when moisture is present.

Epoxide resin

This material is used extensively for 'potting' or encapsulating electronic items. In larger castings it is found as insulating bushings for switchgear and transformers.

Textiles

This group of insulating materials includes both natural (silk, cotton, and jute) and synthetic (nylon, Terylene). They are often found in tape form, for winding-wire coil insulation.

Gases

Air is the most important gas used for insulating purposes. Under certain conditions (humidity and dampness) it will break down. Nitrogen and hydrogen are used in electrical transformers and machines as both insulants and coolants.

Liquids

Mineral oil is the most common insulant in liquid form. Others include carbon tetrachloride, silicone fluids and varnishes. Semi-liquid materials include waxes, bitumens and some synthetic resins. Carbon tetrachloride is found as an arc-quencher in high-voltage cartridge type fuses on overhead lines. Silicone fluids are used in transformers and as dashpot damping liquids. Varnishes are used for thin insulation covering for winding wires in electromagnets. Waxes are generally used for impregnating capacitors and fibres where the operating temperatures are not high. Bitumens are used for filling cable-boxes; some are used in a paint form. Resins of a synthetic nature form the basis of the materials known as 'plastics' (polyethylene, polyvinylchloride, melamine and polystyrene). Natural resins are used in varnishes, and as bonding media for mica and paper sheets hot-pressed to make boards.

36

Workshop measurements

Many kinds of measurement are taken in everyday work in the work-shop: lengths, widths, angles, thicknesses and so on. For some measurements simple tools are available, such as the rule. For others, particularly for accurate work, tools are more complicated (e.g. the micrometer). Each measuring instrument is designed to do a certain job: to provide information in recognisable units for comparison with a group or set of similar units relating to the shape and size of a piece of work. Electrical measurements are also made to discover the electrical pro-perties of materials and the circuits in which they are incorporated. This chapter is no more than a brief summary of the instruments and tools used for measurement work in the workshop.

'Line' and 'end' measurement

It is one thing to check the accuracy of one's work; it is another thing to indicate its size by actual measurement. A length may be expressed as the distance between two lines (called 'line measurement'), or as the distance between two faces (called 'end measurement'). The most common example of line measurement is that which uses the rule, which has divisions shown as lines marked on its surface. For end measure-ment we use calipers, the micrometer, solid length bars and so on to obtain size.

The rule

For workshop measurement work it is best to buy a good rule of the engineer's type. For work outside the workshop, the electrician uses a boxwood rule, because he does not require generally the same degree of accuracy. The most useful and convenient markings on the engineer's rule are inches on one face and metric units on the other. A good rule is worth looking after. It should not be used as a feeler gauge, screwdriver blade or other purpose which may mar its ability to do its specific job. In particular its ends should be protected from wear, because they form one end of a dimension. A rounded corner end could well introduce an inaccuracy of 0·25 mm or 0·5 mm. Rusting of the rule can be prevented by keeping it in an oiled cloth or rustproof paper. When taking measurements with a rule it is best to hold it so that the graduation lines are as near as possible to the faces to be measured. Some rules are provided with bevelled edges, so that 'parallax' is reduced to the minimum. This term is used for the type of error which results from looking at, for example, a clock at an acute angle instead of from the front of it.

Calipers

To measure the diameter of a circular part involves straddling across it to obtain the length of its greatest dimension. The rule is not always the right tool for this, so we use the calipers. The shape of calipers varies according to whether they are to be used for external or internal work. They may be stiff-jointed at the hinge of the legs (opening is maintained by the friction at the joint); or else the joint may be free and spring-controlled (opening is adjusted and maintained by a nut working on a screw). Screw-controlled spring calipers are more easily adjusted. When the calipers are adjusted and indicate the dimension between their legs, the dimension should be read off either on a rule or micrometer. Calipers should be used with care in the workshop and not knocked around so that they become damaged.

The micrometer

When a part has to be measured to, say, the third place after the decimal point in the Imperial system of units, or the second place in the metric system, a more accurate method of measurement is needed, than can be obtained with the use of the rule. The micrometer is used here. It consists of a semi-circular frame with a cylindrical extension (the barrel) at its right end, and a hardened anvil inside, at the left end. The bore of the barrel is screwed with a very fine thread and the spindle, which is attached to a thimble, screws through. The barrel is graduated in divisions of a unit, as is the rim of the thimble. Measurement is taken between the face of the anvil and the end of the spindle. The accuracy of the micrometer depends on the price paid for it. It should be treated with care as it is a most valuable and useful instrument.

The vernier calipers

This measuring device gives an 'end' measurement. The positions of the jaws of the instrument are, however, controlled from a 'line' scale. Transfer of one measurement to the other is made possible by the vernier scale. A vernier scale is the name given to any scale which makes use of the difference between two scales which are nearly, but not quite, alike, so that small differences can be obtained. The vernier is made in various sizes from 150 mm upwards. It is not used for straddling round bars in the same way as a micrometer, but may be used for measuring large diameters on their ends, or large bores.

Gauges

While an instrument is used to measure a dimension, the gauge is used to check the accuracy of a piece of work, without any particular reference to its size. There is an extremely wide range of gauges available, all of which have a particular function in the workshop. Included in the range are *hole gauges*, *limit plug gauges* and *plate gauges*. The material for gauges is extremely hard and resistant to wear. For general electrical work, the wire gauge is used to check the size of a wire. Other gauges in general use are often associated with a manufacturer's product. For example, to check the size of MICS cable, some manufacturers issue a gauge with a large vee-slot and marked off in the actual size of the cable. Where important contracting jobs are being carried out to rigid specifications, it is sometimes necessary to check off, say, the thickness of conduit walls.

Electrical measurements

The measurement of electrical properties of materials and electrical quantities is made by using instruments, the basic principles of which are discussed elsewhere in this book. The properties usually measured are insulation resistance, continuity and conductor resistance. The electrical quantities are current and voltage, generally associated with a circuit in operation (e.g. an electric iron on test). Chapter 18 of this book discusses electrical measurements. Many of the most accurate instruments, particularly the multi-range types, are expensive to buy. But by the time the years of apprenticeship are out, it should be possible for the newly-fledged journeyman with a respect for his work to make the purchase of a small multi-range instrument which, even if the ranges are limited, will still form an extremely useful part of his tool kit. The more expensive instruments are supplied by the employer and their use is generally restricted to experienced men who are able, not only to use them correctly, but to interpret their readings.

37

Workshop practice

Basic workshop operations. A great deal of the activity of the electrician or electrical fitter in the electrical workshop is closely associated with mechanical engineering craft practice. For instance, metals are cut, drilled, tapped and so on. In the context of maintenance and repairs on electrical machines and apparatus, the element of mechanical engineering is often greater than that of the electrical side of things. This chapter is no more than a reasonably comprehensive summary of the techniques involved in some of the more common activities usually found in an electrical workshop. Some of the activities will also be those done by the electrician outside the workshop—on site, in a customer's home and in places where workshop facilities are not available. But, in general, the techniques remain the same.

Riveting

When two flat surfaces (e.g. plates) have to be fastened together to form a permanent joint, riveting is a satisfactory method. Rivets are classed according to the shape of the head. The round snap-head is the most commonly used. But if the projecting head is an inconvenience, the countersunk type enables a flush-head finish to be obtained—though it does not give such an efficient joint. Rivets put in and riveted-up hot give the best results, particularly for heavy work. However, for light work, copper, brass and aluminium rivets may be used cold. For very light work, bifurcated and hollow rivets may be obtained. But these do not give such a satisfactory joint as the solid rivet. Riveted joints to plates or flat bar may be made either by lapping over the edges of the plates and fastening with one or two rows of rivets (lap joint). Or else, the edge of the plates may be butted together and the joint completed by holding them together with one or two cover straps and riveting (butt joints).

When preparing the plates for the joint they should, if possible, be clamped together with the top plate marked out for the holes. The holes are then drilled in all the plates at once. Using this method, there will thus be no doubt about all the holes being in exact alignment when the rivets are put in them. If one plate has already been drilled, it should be clamped in position and the holes marked through for drilling in the other plates; or the holes can be used as guides for the drill itself. The riveting process, either using heat or while cold, is the closing of the rivet: supporting the rivet head while the plain end is riveted over using an appropriate cupped punch or the ball-end of a hammer. The process is to swell the metal in the hole to fill it and so pull the plates tightly together. Care should be taken to see that the rivet end is spread evenly in all directions, and not bent over one way. Countersunk rivets require finishing with the flat end of a hammer.

Soldering

Soft soldering is a quick and useful method for making joints in light articles made from steel, copper, brass and aluminium, and for the conductor joints which occur in electrical work. Soldering itself does not make a strong joint. Mechanical strength is obtained by, say, marrying conductor strands or bolting busbars, and then soldering. Where joints are required to be mechanically strong, they should be riveted, brazed or welded. The subject of soldering has been dealt with in Chapter 38.

Silver soldering is a hard soldering process which falls between soft soldering and brazing. There are two common silver solders:

Grade A: silver 61 per cent; copper 29 per cent; zinc 10 per cent. The melting range is 690°C–735°C.

Grade B: silver 43 per cent; copper 37 per cent; zinc 20 per cent. The melting range is 700°C–775°C.

A good soldered joint is recognised by a small amount of solder and perfect adhesion—rather than large unsightly blobs of solder. The following hints may be found useful:

1. Always use a bit with plenty of heat capacity.
2. A better joint can be made if the joint is warm rather than cold.
3. Iron tinning is made easier by having some blobs of solder in a tin lid with a little spirits, and touching both the spirits and the solder at the same time.

Brazing

The joining metal here is brass, which is a harder, stronger and more rigid metal than solder. As in soldering, an alloy of the brazing metal and the metal of the joint is formed at the surface of the joint metal.

The brass used for making the joint in brazing is usually called 'spelter'. Its composition depends on the metal being brazed, because it is essential that the spelter has a lower melting point than that of the material to be joined. Spelter is obtainable in sticks or in a granular state, when it may be mixed with the flux (borax) before being applied to the joint. The heating for the joining process is obtained from a blow-pipe or blow-torch. The rules of cleanliness which apply to soldering alloy also apply to brazing; the work should be cleaned and polished at all points where the brazing is to occur. The work should be allowed to cool off normally, as quenching may lead to distortion of the joint or cracking of the spelter.

Chiseling

For a description of the chisel see Chapter 32. The cold chisel is an important cutting tool. An engineer's chisel does not have a wooden handle as do other types (e.g. woodworking). In the cutting process using a chisel, the tool should always be held about half way between the head and the edge, and at the correct inclination. If the angle is too great, the edge will cut too deeply. On the other hand, a shallow angle will not allow the chisel to cut efficiently. Generally, the smaller the angle the more effective are the hammer blows, and the more efficient the cutting operation. The edge should be kept well up against the shoulder formed by the cut and chip. Particles of metal should be kept away. When the chisel approaches the edge of the metal to be cut, particularly cast-iron, it should be reversed or the cut taken at right-angles to the previous one. Otherwise the edge of the metal is likely to be broken away.

Cutting with the hack-saw

The hack-saw is the chief tool used by the fitter for cutting-off, and for making thin cuts before chipping and filing operations. The choice of blade for any class of work is governed mainly according to the pitch of the teeth on the blade. These should be as large as possible to give the greatest clearance for the metal chips and to avoid clogging. Two at least, or three, teeth should always be in contact with the surface being sawn. Otherwise there will be a danger of the teeth being stripped from the blade. Also, if a corner is sawn too sharply, the teeth may get ripped off. Blades must be held tightly in the frame. Slow, firm and steady strokes (about 50 per minute) are best. On the return stroke the blade should be lifted slightly. Breakage of blades is caused by (*a*) rapid and erratic strokes; (*b*) too much pressure; (*c*) the blade being held loosely in the frame; (*d*) binding of the blade from uneven cutting; (*e*) the work not being held firmly in the vice. Solid metals should be cut with a good firm pressure; thin sheets and tubes need a light pressure. If the pressure at the start of a cut is insufficient, the teeth may glaze the work, and so cause their edges to be rubbed away.

Filing

To produce a flat surface by cross-filing is a difficult task: practice makes perfect in this activity as in others, and advice from skilled filers should not be ignored. The work should be held firmly in the vice, in a truly horizontal position. Grasp the file handle with the right hand, with the end of the file-handle pressing against the palm of the hand in line with the wrist-joint. The left hand is used to apply pressure at the end of the file. Strokes should be made by a slight movement of the right arm from the shoulder, and by a sway of the body towards the work. To get the movement right it is necessary to take up a stance to the left side of the vice, with the feet planted slightly apart. As the file moves over the work, it should be in an oblique direction, rather than parallel to its length. The file must remain horizontal throughout the stroke, which should be long, slow, and steady. Pressure is applied only on the forward motion. On the return stroke, although the pressure is relieved, the file remains in contact with the work.

Success in filing flat is dependent upon keeping the file really horizontal throughout its stroke. This position is controlled by the distribution of pressure between both hands. The fault with beginners is that too much pressure is applied on one hand, resulting in a round rather than a flat surface. Any tendency to rock should be corrected immediately. To test the surface of the work during filing, use a straight edge during the process and view the line of contact for 'daylight'. If a considerable amount of metal has to be removed, the bulk of the metal should be taken away by using a rough file. The surface should be progressively brought to a finish by using second-cut and smooth-files.

Draw-filing is the process used to remove file marks and to give a good finish to the work. For this purpose use a good fine-cut file with a flat face (e.g. a mill file).

Scraping

The purpose of the scraper is to correct slight irregularities from flatness. If these are great, a file should be used, for the scraping process is rather laborious and not intended for removing much metal. First, a standard of flatness must be obtained with which to compare the surface being scraped. A surface plate is used for this comparison. After thoroughly cleaning the surface plate, it should be smeared with a thin layer of prussian blue, or red lead in oil. The other surface (the work) should then be placed on it and rubbed about slightly. The high parts of the work surfaces will be smeared with part of the marking substance. These are the parts that must now be reduced by scraping. This is done by holding the scraper firmly to make short strokes 25 mm long. The scraper is left in contact with the surface for the return strokes, though no pressure is applied. Having gone over the surface with strokes in one direction, the next set of strokes should be made in the other direction, at right-angles to the first. This changing round procedure is repeated until sufficient metal is removed. From time to time trial rubbings should be made with the surface plate, and working on those parts which show 'high'. Gradually the surface of the work is brought to the condition that its whole area is covered by tiny areas of contact; it is usually taken as accurate if these spots are 3 mm to 5 mm apart. The scraper must always be kept sharp.

Drilling

If the work to be drilled is large and heavy there will generally be no danger of it moving or rotating dangerously, with the movement of the drill. Otherwise, with light work, always clamp or hold it securely by some method. Secure clamping ensures no danger to the drill operator, more accurate drilling and less breakage of drills. The chief danger in drilling is when the drill point breaks through at the underside of the part being drilled. When feeding the drill by hand, pressure should be eased off when the drill point is felt to be breaking through. Small drills should always be fed by hand. Special care is necessary when drilling thin plate because the drill point often breaks through before the drill begins to cut on its full diameter. The following hints are useful:

1. For soft metals use a drill with a quick twist to its flutes. The opposite applies for hard metals. For chilled iron, a flat drill gives the best results.
2. Cut with soluble oil for steel and malleable iron, and kerosene or turpentine for very hard steel. Cast-brass or iron should be drilled dry.
3. The blueing of a high-speed steel drill is not detrimental to it; but it is to a carbon-steel drill.

4. When drilling deep holes, release the feed occasionally and withdraw the drill. Remove metal chips from the bottom of the hole with an old round file that has been magnetised.

Marking out

Unless a particular job is simple, it is usual to indicate with marks where cuts, holes, slots and so on have to be carried out. The process of marking out means to use certain tools to prepare the job for some subsequent process to produce a finished piece of work.

The *marking out table* has a metal (steel) top which is planed dead flat and is supported firmly. The *angle plate* is used for supporting a surface at right-angles to the surface of the table. It has plenty of holes and slots for taking the bolts necessary to secure pieces to it. The *adjustable angle plate* is used for angular work. *Vee blocks* are used to support round bars while they are being marked off. They are usually made in pairs. *Parallel strips* are used for supporting work on the marking-out table. Each pair of strips is of exactly similar dimensions, with parallel faces and all faces square. The *spirit level* is used for setting off surfaces parallel to the surface of the marking off table. The level is also used for levelling machine beds and tables. A pair of *toolmaker's clamps* is an essential item in the fitter's kit; two or three pairs are the minimum number. They are handy items for clamping work, say, to an angle plate. *Dividers* and *trammels* are used for scribing circles, marking off lengths and so on. Trammels are used to extend beyond the range of dividers. Trammels with bent legs are sometimes used in the same way as inside calipers.

Feeler gauges consist of a series of blades or leaves of thicknesses varying from about 0·002 mm to 1 mm. They are useful for flatness tests. Other tools for marking off with accuracy include the *try square* for testing squareness, *protractors* and the *bevel* for testing the angle between two surfaces.

For any scribed line to be visible it is first necessary to prepare the surface which is to receive the line. The rough faces of castings which have to be marked should be brushed over with a little whitewash which, when dry, will show up the scribed line. Machined surfaces may be prepared by brushing them over with copper sulphate solution which leaves a thin film of copper on the surface; this shows up a scribed line very clearly. It should be noted that marking out does not mean that one can dispense with the use of measuring instruments for the control of machining and cutting processes. Always the line should be used as a guide, the final finishing being done to micrometer, vernier, or some other exact measuring method. Chapter 36 deals with workshop measurements.

Drilling and tapping a hole

For tapping it is first necessary to drill a hole the diameter of which must be approximately equal to that of the bottom of the thread. When the hole to be tapped is 'blind' (i.e. open at one end only) it should be drilled one or two threads deeper than is required for the finished depth of the tapped hole.

Having drilled the tapping hole, the taper tap is secured in the tap wrench and started in the hole. Before beginning to screw it round for cutting the thread, its position must be adjusted until it stands square with the top surface of the work. This can be done by setting the tap vertical with the aid of a small square. Enter the tap and keep it straight for the first few turns. A little oil will help the action of the tap and improve the finish of the threads. When the taper tap is felt to have started cutting, and its squareness checked, the cutting of the thread proper may proceed. The tap should not be turned continuously, but it should be reversed about every half-turn to clear the threads of metal particles. If any stiffness is encountered (this can be felt on the tap wrench), no force should be used, but the tap carefully eased back and fore to clear the obstruction. When a blind hole is being tapped, the tap should be withdrawn from time to time and the metal cleared from the bottom of the hole. If the hole is a straight-through one, the reduction in resistance will indicate that the taper tap is cutting the full thread. It can then be withdrawn and replaced with a second tap. When a blind hole is being tapped, increase in resistance will be felt as the point of the tap reaches the bottom of the hole. No force must be used at this point as the tap may be broken or the threads stripped. Remove the taper tap and take a second tap down as far as it will go. Remove this tap and replace it with a plug tap. Great care should be exercised when using small, slender taps, particularly in blind holes, as the tap may break inside the hole. If this happens, the piece may be removed with a punch. Otherwise, it will have to be softened by heating, drilled out and the hole retapped.

Threading

The tool used for cutting external threads on bars and tubes is called a die. Basically it consists of a nut with portions of its thread circumference cut away and shaped to provide cutting edges to the remaining portions of the thread. The die is used with a pair of operating handles called stocks. The action of dieing is similar to that of tapping, except that it is more difficult to keep a die square. It is therefore necessary to exercise great care in threading. When the cut has begun, the die should be worked back and fore similar to the method used for tapping. Certain designs of stocks and dies incorporate guides which ensure that the die is kept square-on. The conduit stocks and dies for the electric thread show an example of this provision.

Grinding

This process involves the use of abrasive wheels to cut, surface and shape materials. The most common type of wheel is the straight grinder, and is generally used in the workshop for sharpening tools and the rough shaping of a surface before a final finishing process. The main points to bear in mind include the wearing of goggles to protect the eyes, and the holding of the work piece securely.

Fitting pulleys to motors

First ensure that both the motor shaft and the pulley bore are quite clean. The key should be fitted into the shaft, taking care to see that there is clearance between the top surface of the key and the bottom surface of the pulley keyway when the pulley is in its final position. Thus, the key should fit only on the sides of the keyway. The shaft and bore should then be lightly smeared with lubricating oil and the pulley worked onto the motor shaft. Ensure that the shaft key is in line with the pulley keyway. The pulley should be driven home by light taps with a hammer on a piece of wood held against the pulley hub. The blows of the hammer should be distributed round the whole surface of the hub to ensure uniform pressure. The pulley should be driven right home and any set-screw firmly tightened onto the key.

Extraction of pulley

This task is best done with the use of the tool called 'pulley drawers'. It consists of a pair of jaws coupled together at one end and containing a long threaded bolt which, when turned, exerts pressure on the motor shaft and so draws off the pulley. A small piece of copper or other soft metal should be placed between the point of the screw and the shaft end to avoid unnecessary damage.

Overhauling a motor

The following is a summary of the procedure to be followed when overhauling a motor in the electrical workshop:

1. Dismantle the motor carefully; don't open cartridge bearing housings.
2. Clean away all dust, dirt, oil and grit by using a blower, compressed-air hose, bellows or brushes, and with petrol if necessary, or carbon tetrachloride (CCl_4). The complete removal of any foreign matter is important.
3. Check all parts for damage or wear, and repair or replace as found necessary.
4. Measure the insulation resistance and dry out windings if necessary until the correct value is obtained. Replace or repair any damaged windings.
5. Reassemble all parts. Ensure that machine leads are on the correct terminals and that all parts are well tightened and locked, where this provision is made.
6. Check insulation resistance.
7. Check air gaps.
8. Commission and test.

38

Soldering

The process of soldering in electrical work is generally used to make joints in conductors and to terminate a conductor for a mechanical connection to an appropriate terminal in electrical equipment. With the emphasis nowadays on speed, while still producing a good job of work, many aids to jointing have been produced. For instance the married joint which was once used to make a straight-through connection between the ends of two cables is now being replaced by the weak-back ferrule. And the tee joint is being replaced with the use of the claw-type clamp.

The bulk of information relating to conductor joints and terminations is contained in Chapter 7 of this book. This chapter deals with the subject of soldering as a workshop practice, and actual jointing procedure is only touched on where relevant.

The nature of soldered joints
The essential feature of a soldered joint is that each of the joined surfaces is wetted by a film of solder, and that the two films of solder are continuous with the solder filling the space between them. When solid, the joint becomes a hard mass of the material to be joined and the solder.

Basically, solders are metallic substances which have lower melting points than the metals they are to join together. They act (a) by flowing between the metal surfaces to be joined; (b) by filling completely the space between the surfaces; (c) by adhering to the surfaces; and (d) by solidifying. Once it adheres to a metal surface, solder cannot be prised off, because it becomes attached chemically to the surface and forms an intermediate compound. Thus, the soldered joint is, if made correctly, a most effective method of producing an electrically and mechanically sound form of connection between two current-carrying conductors.

The basic steps in soldering
Making a soldered joint can be divided into three basic steps:

1. Shaping the metal parts so that they fit closely together.
2. Cleaning the surfaces to be joined with a special substance.
3. Applying the soldering flux.
4. Applying molten solder.
5. Removing surplus solder and cooling the joint.

The surfaces to be joined should fit so that the space between them is narrow enough to become completely filled with molten solder (drawn in by capillary action as the solder wets the metal). If there is not enough clearance between the surfaces, the solder may not be able to penetrate and the result may be a joint with holes.

The reason for cleaning the surface is to expose the bare metal and make it free from grease or oxide which would otherwise prevent the solder from adhering to the metal. Rough cleaning consists in filing, scraping or rubbing with an abrasive cloth or paper. However, to get the surfaces really clean it is necessary to apply some substance which will remove any remaining particles of foreign matter. For large jobs degreasing is done by a solvent application or by cleaning in a bath of weak solution of an alkali (sodium salts) in hot water. In smaller jobs, such as jointing the smaller sizes of conductor, the flux used is sufficient to clean the surfaces of the metals to be joined together.

When the metal surfaces are being heated prior to the soldering operation, oxides may form on them. To prevent this happening a flux is applied. The basic characteristics of any flux are:

(*a*) It should be a liquid cover over the metal and exclude all air.
(*b*) It should function as (*a*) while the surfaces are being heated up to soldering temperature.
(*c*) It should dissolve any oxide that may have formed on the metal surfaces, or on the solder itself, and carry these away.
(*d*) It should be easily displaced from the surfaces once the molten solder is applied.

Fluxes for electrical work must not remain acidic or corrosive at the completion of the soldering operation. Thus, acid fluxes such as 'killed spirits' should not be used, because they tend to cause corrosion in the joint after soldering. The fluxes used in electrical work include pure amber resin (or rosin) in some form, 'activated' resin as used in cored solders, and solder paste. Resin is the gum exuded from artificially-produced wounds in the bark of pine trees. At ordinary temperatures it is solid and does not cause corrosion, but it reacts mildly at close to soldering temperatures. Resin is easily crushed to a powder form and melts readily at 125°C. It can be applied by sprinkling the powder on the joint, but the more effective method is to brush it on to the joint as a solution. The usual solvent for resin is methylated spirit or industrial spirit. Activated resin fluxes contain a chemical which reacts with the oxides on the heated surface of a metal to clear them away, but they do not remain corrosive after the soldering operation.

Solder pastes are used where liquid fluxes would drain away too quickly from the surfaces to be joined.

The next step in the soldering process is the tinning which, in the case of tinned-copper conductors, has been done already. This process is the spreading of a thin layer of solder over the surfaces to be joined so that the surfaces are 'wet'. If the solder does not spread quickly and easily, then this indicates some foreign matter in the area (either oxide or grease). If this is the case, the surfaces should be cleaned with the application of a fluxed cloth-pad, more flux and then retinning. Without the wetting of the metal surfaces there can be no soldering action. The molten solder should leave a continuous permanent film on the surfaces instead of rolling over them.

Once the joint surfaces are thoroughly wetted and the space between them is completely full of solder, it remains only to cool the joint after wiping off any surplus solder using the fluxed cloth-pad. Cooling should in most instances be natural. Rapid cooling may cause a 'dry' joint which has a high resistance—the result of the soldering cracking.

During the soldering process it is essential to ensure that the job is clamped securely and that the clamping device holds the parts to be joined firmly and accurately.

The soldering bit

The function of the soldering bit is to store and carry heat from the heat source to the work, to store molten solder, to deliver the molten solder and to withdraw surplus molten solder. As a carrier of heat the bit should have a large heat-storage capacity. The size of the bit for any soldering job is determined by the rate at which it has to supply heat to the work. The material for the bit is most often copper, which wets readily and to which molten solder clings without difficulty.

Various methods are used for heating the bit. In days gone by the source of heat was a stove burning charcoal or coke: coal fires were not favoured because of soot. Nowadays coal-gas, paraffin or petrol flames and electric heating are used.

Care of soldering bits

Soldering bits rapidly become coated with scale (oxide) and the faces that are wetted by flux or solder become pitted. Pitting develops during the life of the bit, sometimes quite early in its life. The result of a pitted bit is that the work becomes impaired and it becomes necessary to file the faces of the bit until a sufficiently large surface area is obtained and tinning becomes easy. Pitting is the result of (a) oxidation of the copper due to heat, (b) attack by the flux on the copper face and (c) the transfer of copper particles from the bit into the solder. Care involves the maintenance of the soldering temperatures and no higher. Excessive temperatures result in whole-scale oxidation and rapid wear. At all times the shape of the bit should be preserved. Each bit should be of a shape appropriate to the job.

Wiped joints

A wiped joint is a full-bore straight connection between two pipe-like sections. In electrical work one section is the lead sheath of a cable; the other section is the entry to a cable sealing box, or a straight-through or tee-joint box. Making a wiped joint is a skilled task and is usually performed by experienced cable jointers.

There are two methods:

(a) The first is the blowlamp method, though a blow-torch is more often used. The area to be jointed is heated by playing the flame of the torch over it. A solder stick is held in the flame and, when the surface is hot enough, is applied round the joint in order to tin it. The solder is rubbed over the joint area which has been previously prepared with flux (tallow is often used). Additional solder is melted onto the area and is caught on a wiping cloth made from several layers of buckskin cloth. The molten solder is worked round and round to ensure an even consistency in the plastic mass. When enough solder has been applied, the solder stick is laid aside and the mass of plastic solder melted off onto the cloth. The cloth, of course, must always be kept smeared with tallow and heated on its working face.

The mass of the solder is then lifted back round the joint and wiped into shape without the flame. The molten pasty solder must be kept moving and the shaping completed before any indications of setting are seen. This is skilled work which requires experience to do a good job. This method is suitable for horizontal working.

(b) The second method of making a wiped joint is suitable for vertical joints and is known as the pot-wipe method. One advantage of this method is the reduced risk of overheating the lead sheath. The joint is heated by dribbling molten solder over it from a ladle. The solder is melted in a pot. At first the solder is chilled by the coldness of the joint area and solidifies around it without adhering. As the pouring is continued the mass gathers heat and becomes pasty. The pipe surface is raised in temperature and becomes tinned as the solder melts off the pipe to be caught in the wiping cloth. The wiped joint is formed as already described. A circular catch-plate of sheet metal with a central hole is fixed round the lead sheath. To prevent solder from adhering to the sheath beyond the required limit on each side of a wiped joint, plumbers' black is used and is applied beforehand.

About solders

Two basic types of solder are used in electrical work: fine solder (tinman's solder) which is 60 parts tin and 40 parts lead, and plumber's metal which is 30 parts tin and 70 parts lead. Fine solder melts more easily and so is more commonly used for electrical joints. Fine solder is used for conductors; plumber's metal is used for wiped joints because it remains in a plastic state longer than the finer metal.

Solder is made in a variety of forms: bars and sticks, and in wire and cored forms. Wire solder is convenient to use and comes in various diameters. For fine electrical and electronic work finer gauges of wire are available. These wires melt instantly at a touch of the copper bit and their uniform gauge makes it easy for just the right amount of solder to be applied to a joint. Cored solders are hollow or grooved wires or tubes filled with flux. The flux is introduced warm as a paste during the drawing of the tube, and solidifies when cold. With the flux in the cored solder some considerable time is saved when a number of small joints have to be done quickly. Solders are also available in paste, cream and powder forms.

Soldering aluminium

For many years it has been accepted that aluminium cannot be soldered easily or satisfactorily by methods commonly used for soldering copper. The oxide film which forms with heat on the surface of aluminium is very resistant to chemical attack, and although the film can be removed by mechanical means, the cleaned surface thus exposed will oxidise very rapidly and the abrasive method of cleaning is, therefore, a very uncertain one. This problem has been solved by the development of new fluxes which will cope with this oxide film and it is now possible to solder aluminium by the conventional methods used in soldering other non-ferrous metals. The fluxes are available in paste and liquid forms and incorporated in cored solder-wire. The solders are in stick and ingot forms.

Soldering aluminium today is now a straightforward process. The following points must be observed:

(*a*) All surfaces must be scrupulously clean.

(*b*) When a joint is being made between stranded conductors the strands must be 'stepped' to increase the surface area.

(*c*) The surface must be heated before the flux is applied. The flux will take only when the temperature is high enough.

(*d*) Apply aluminium solder until the surface is bright.

(*e*) Joints in aluminium should always be protected from contact with the atmosphere: by painting, taping or compounding.

Part C

Appendices

Appendix A

Typical report on an ac electric motor

MACHINE NO: P29		MAKER'S NO: TS 13567		
TYPE: Slip-ring induction		ENCLOSURE: Totally-enclosed, fan-cooled		
kW(cmr): 30	VOLTS: 415	PHASES: 3		Hz: 50
RPM: 960		F.L. AMPS: 44		
DRIVE: VEE		MOUNTING: Foot		BEARINGS: Roller
FUNCTION: Vee-pulley drive for rack circular saw bench				

Note. The following is a typical routine service report for a fictitious ac motor. Certain aspects of it pertain to a squirrel-cage motor; others pertain to a slip-ring motor. The aspects have been combined to form the basis of what a typical report should generally contain. It is assumed initially that the performance of the motor has been satisfactory (verified orally by the machine operator) and the present report is to confirm in reasonable detail the electrical and mechanical condition of the machine together with its control and starting equipment. The information which is obtained as a result of the tests and checks carried out should be brief and concise. Comments on the state of the machine and its associated control gear should be incorporated. If further action is thought necessary at the next service check, a detailed statement of the necessary points to be acted upon should be the subject of an appendix.

Motor: Electrical

(a) *Stator winding*

1. Insulation resistance of windings tested with a 500V Megger. Atmosphere warm and dry. Value obtained: $7\frac{1}{2}$ Megohms.
2. Continuity of windings checked out with Bridge-Megger. Continuity satisfactory. Resistance of windings balanced and checked against maker's original figures.
3. Visual inspection of the windings' insulation revealed no defects nor potential faults due to overheating.

(b) *Rotor winding*

1. Brushes lifted off slip-rings and insulation resistance of windings tested with a 500V Megger. Atmosphere warm and dry. Value of *IR* obtained: $8\frac{1}{4}$ Megohms.

2. Continuity of windings checked out with Bridge-Megger. Continuity satisfactory. Resistance of windings balanced and checked against Maker's original figures.

3. Visual inspection of winding insulation revealed no defects nor potential faults due to overheating.

4. Slip-rings examined. Surfaces of rings in excellent condition. Brushes are bedded correctly with 3 Kg pressure (checked with spring-balance). Area of slip-rings clear of oil and dust. Insulation surfaces clear of foreign matter.

5. All electrical connections to brushes and holders are in good condition. Brushes move freely in holders. Conductors from slip-rings to rotor windings in good condition.

(c) Circuit conductors

1. In terminal box one locknut found slack. This was tightened. All other nuts, washers, lockwashers and locknuts checked and found to be tight. Slight deposit of wood dust on terminal-block surface was wiped off. Cover of terminal box examined and groove packing (found dry) recharged with grease. Seating of cover on box checked on replacement and tightened up securely.

2. Insulation resistance of conductors to earth and between conductors tested with 500V Megger. Readings: infinity between all conductors; 20 Megohms between R and B phase conductors and earth; 12 Megohms between Y phase conductor and earth.

3. Continuity of ECC between motor-starter and motor found to be satisfactory at 0·003 ohm. Earthing terminations visually inspected and found tight, secure and protected from possible mechanical damage. No sign of corrosion at copper-steel contact.

Motor: Mechanical

1. Motor frame checked and found to be mechanically sound. Air-gap checked with feeler-gauge round periphery of rotor and found to be satisfactory. From this check it was found that there was a difference of 0·025 mm, indicating bearing wear. Recommend replacement of bearings after an estimated 2,000 hours running. Bearings cleaned (diesel oil) and relubricated with Shell 'Alvania 3'. General cleaning of motor casing carried out. No cleaning of interior of motor was required.

2. Vee-pulley checked for correct alignment and found to be satisfactory. Vee-ropes now showing signs of wear and slipping in vee-grooves of pulley. Slight application of belt syrup carried out.

Control Gear

1. All main and auxiliary contacts inspected visually and found healthy. Settings of overload trips checked. Time-delay dashpots checked for correct oil-level. Contactor mechanism found dusty. Cleaned thoroughly and lubricated at pivot points. Traces of wood particles found on face of coil magnet (operator of machine verbally reported 'chattering' in starter). Particles

removed and contactor mechanism bedded down onto magnet face. Contactor small wiring tested for insulation-resistance to earth and between conductors. Satisfactory readings obtained: both Infinity. Ammeter connections checked and found tight and secure.

2. Isolator checked. Switch-blade on Y phase conductor terminal (motor side) making bad contact. Visual inspection showed signs of burning, slight pitting and overheating. Contact cleaned up and restored to good order. All conductor connections and terminations are now tight. Mechanical operation of isolator found to be sluggish. Small amount of lubricating grease (petroleum jelly) applied. Operation now satisfactory.

General

The motor and its associated control and starting gear is in a very good condition. Low value of IR-to-earth on Y phase conductor may be due to faulty switch-blade contact found in isolator with resultant deterioration of insulation. Recommend further check after 120 hours' running (two weeks). If the value is still low, suggest replacement of conductor. Vee-ropes require weekly check from the date of this report.

After this inspection, the supply was restored to the motor.

Motor started up and placed on full load. Temperature rise after 1 hour on full load was 2 °C (measured by thermometer).

DATE OF TEST AND INSPECTION: 27 June 1978
DATE OF THIS REPORT: 29 June 1978
SIGNED: J. Brown, Asst Engineer, Electrical Dept

Appendix B

Consolidation questions

The following questions are not generally Examination Questions. They are designed rather to consolidate the material given in the chapters of this book. It is hoped that students tackling the questions will endeavour to find the full answers to questions from as many sources of information to which they can find access. In this way, their general knowledge of the subject of electrical installation technology and its related subjects will broaden to their own benefit.

1. Describe how to resharpen a twist drill: mention any precautions to be observed.
2. Why are the copper strands of conductors usually tinned?
3. List the important precautions to be observed when preparing a cable termination.
4. Describe the features of the various types of conductor joints.
5. Why is a flux necessary when soldering?
6. List the grades of solder used in electrical work.
7. Describe a practical method of determining the point at which the solder in a metal pot has reached its correct working temperature.
8. Compare the compressed and solder methods of cable joints and termination.
9. Why is it necessary to double back the conductor when terminating a single wire?
10. What is the difference between a flexible cable and a flexible cord?
11. Describe one form of cord grip or cord anchorage.
12. What is the function of the lead-alloy sheath in the LAS wiring system?
13. Why are the fuses in a consumer unit not interchangeable outside their own group?
14. Compare the three-plate ceiling rose with the two-plate rose.
15. Compare the use of a backplate lampholder and a flexible lighting pendant.
16. Why are some switches marked 'For ac only'?
17. Compare the following terms: surface, semi-recessed and flush.
18. Why is the term 'trembler' used with this type of electric bell?
19. Describe the operation of a continuous-ringing electric bell.
20. List as many as possible uses for the electric bell.
21. List the IEE Regulations requirements regarding ELV bell circuits.
22. Why must the windings of a bell-transformer be double-wound?
23. What is the purpose of a bell-indicator board?
24. Measure the resistance of a lamp filament when cold. Calculate its resistance when hot. Why is there a difference in resistance values?

25. List the full requirements of the IEE Regulations regarding the use of the cleated wiring system.
26. Explain why electrical continuity is important in a conduit installation.
27. Describe a method of testing the electrical continuity of conduit.
28. Describe fully the methods used for fixing conduits.
29. Describe a method used for securing a conduit switch-drop to a chase in a brick wall.
30. How does the 'between conductors' test for *IR* differ from the 'between conductors and earth' test?
31. List the types of installation metalwork which are not required to be earthed.
32. List some locations in a building where two-way and intermediate lighting circuits may be installed with advantage.
33. Comment on the use of black-coloured wires as switch-wires.
34. Explain fully the purpose of a fuse.
35. Explain the purpose of an isolating switch.
36. Explain fully the purpose of a circuit-breaker.
37. In soldering cable connections, certain precautions must be observed to prevent damage to conductors and insulation. Describe these precautions.
38. Why are sealing boxes necessary when terminating paper-insulated cables?
39. Under what conditions can 250V cables be used in three-phase circuits?
40. Is it permitted that neutral conductors from separate subcircuits be 'looped' together at distribution boards?
41. What are the regulations' requirements for non-metallic conduits with non-screwed joints?
42. What types of cables are not suitable for drawing into concrete ducts?
43. What is the cause of condensation in conduits?
44. What are the IEE Regulations' requirements with regard to metallic trunking joints and junctions?
45. Give a diagram of connections of a voltage-operated earth-leakage circuit-breaker.
46. Explain the operation of a voltage-operated earth-leakage circuit-breaker.
47. Define 'earth-continuity conductor'.
48 Define 'earth electrode'.
49. Define 'earthing lead'.
50. Discuss the connection of an earthing lead to a main water pipe, the pipe to act as an earth electrode.
51. Explain the term 'insulation resistance' of an installation.
52. Explain how the insulation resistance of a piece of insulated cable differs from its conductor resistance.
53. Describe the precautions necessary while erecting a metallic (steel) conduit installation.
54. What is the cause of a fuse blowing when no load is connected?
55. What is one cause of a fuse blowing when connected to a load?
56. What is the cause of reduced voltage at the terminals of a current-using appliance?
57. What is one cause of shocks from the framework of apparatus?

58. Give two examples of conditions where special enclosures are required to protect a motor; indicate the type of enclosure you would select.
59. A 150 cm hot-cathode fluorescent lamp is rated by the makers at 80W. For the purpose of calculating the load of an installation, what is the figure you would use, and why?
60. In what conditions are fireman's emergency switches required?
61. What precautions must be taken when two dissimilar metals are joined together?
62. What instruments are required to test the condition of a secondary cell?
63. What is meant by phase voltage?
64. What is a dc three-wire system?
65. What is meant by the torque of a motor?
66. What is meant by the rating of a motor?
67. What is a 'shunt characteristic'?
68. What is a 'series characteristic'?
69. What is meant by the 'time rating' of a motor.
70. How is the insulation-resistance of a motor winding checked?
71. If the insulation resistance of a motor winding is low, what is the most likely cause?
72. What is the purpose of a motor enclosure?
73. What is the effect of dust on the windings of a motor?
74. What are the two main kinds of induction motor?
75. How is a squirrel-cage rotor constructed?
76. What insulation is required in a motor?
77. What are the main functions of control gear for motors?
78. What devices are required to give adequate protection to a motor?
79. What is direct-on-line starting?
80. What methods are employed to reduce the starting current of squirrel-cage induction motors?
81. When are single-phase motors used?
82. What types of single-phase motors are in use today?
83. What is the principle in operation of the single-phase induction motor?
84. What is a capacitor-start motor?
85. What type of capacitor is used for a capacitor-start motor?
86. What is a 'universal' motor?
87. How are universal motors started?
88. What are the applications of universal motors?
89. What are the two basic types of dc motor?
90. What are the basic components in dc motor construction?
91. Why is it necessary to have special starting gear for dc motors?
92. In what circumstances may dc motors be switched directly onto the supply?
93. What is the usual method of starting a dc motor?
94. How is overcurrent protection provided on a dc motor?
95. How is undervoltage protection provided on a dc motor?
96. Compare the lead-acid cell with the alkaline for modern use today.
97. List the points of maintenance required by a dc motor.
98. List the points of maintenance required by a squirrel-cage motor.

99. List the points of maintenance required by a wound-rotor motor.
100. List the points of maintenance required by a motor starter.
101. List the points of maintenance required by an oil-cooled transformer.
102. What type of motor and control gear would you install for a motor operating in a weather-exposed situation?.
103. Explain what is meant by 'volt drop' and say how it can be avoided.
104. Discuss the provision of armouring for cables.
105. What wiring system would you use for a farm installation?
106. What wiring system would you use for a petrol store?
107. What wiring system would you use for a boiler-house?
108. What is the advantage of using a ring-main circuit with fused plugs?
109. What type of electrical equipment would you recommend for a dairy-house?
110. Discuss the general mass of earth as an electrical conductor.
111. List the tests required for an electrical installation.
112. Discuss the need for Wiring Regulations.
113. Discuss the need for Statutory wiring regulations.
114. Discuss the social aspect of electricity today.
115. Discuss the main requirements of a good lighting scheme.
116. Why is a choke necessary in an electric discharge-lamp circuit?
117. Compare tungsten-filament lighting with fluorescent lighting.
118. What appliances would you recommend for heating a village hall?
119. What appliances would you recommend for heating a bedroom?
120. What appliances would you recommend for heating a small workshop?
121. What are the three methods of heating water by electricity?
122. Describe one form of radiant heater.
123. Describe one form of convector heater.
124. What are the three main parts of a cable?
125. Why is a weight limit placed on a flexible cord?
126. Why are colours used in identifying cables and conductors?
127. What is the difference between a joint box and a junction box?
128. How is an LAS system of wiring adapted to the conduit system?
129. Why must joints in conductors and cables always be accessible?
130. Why must a joint be mechanically sound before soldering?
131. What is the advantage of the 'loop-in' system of wiring?
132. What are the limitations in the use of three-plate ceiling roses?
133. Define a 'final subcircuit' and illustrate your answer.
134. What is the difference between a 'switch feed' and a 'switch wire'?
135. Why are limits placed on the bending of cables?
136. Why are limits in the working temperature placed on cables?
137. Compare the conduit system with the trunking system.
138. Why must a conduit system be erected before wires are drawn in?
139. Why is it good practice to use inspection boxes of circular type rather than channel-type inspection fittings?
140. Why must cables be ventilated?
141. Why is a fire barrier required for busbar trunking?
142. Where are vertical rising-mains used?
143. What are the advantages and disadvantages of the rewireable fuse?

144. What are the advantages and disadvantages of the HRC fuse?
145. What are the advantages and disadvantages of the cartridge fuse?
146. What are the advantages and disadvantages of the miniature circuit-breaker?
147. What is meant by the current rating of a fuse?
148. What is meant by the current rating of a cable?
149. What are the basic requirements for earthing as required by the IEE Regulations?
150. What is meant by Protective Multiple Earthing?
151. Why is a polarity test important?
152. What are the main causes for low insulation resistance?
153. Why is there a difference in the *IR* values expected of an installation when it is complete and incomplete?
154. Why is there a difference in the *IR* value expected of an installation when it is wired with VRI and PVC conductors?
155. Why is it necessary to test the ECC?
156. At what intervals should an installation be inspected? Why?
157. What is the purpose of the power-factor capacitor in a fluorescent lamp circuit?
158. Describe the operation of a switch-start fluorescent-lamp circuit.
159. Describe the operation of a glow-start fluorescent-lamp circuit.
160. Describe the operation of the instant-start fluorescent-lamp circuit.
161. What is the function of a bell indicator?
162. Why is a bell transformer double-wound?
163. Why must all the wires of conductors be securely anchored?
164. What should be done when a cable insulation is damaged by soldering?
165. What types of wiring and equipment are not covered by the IEE Regulations?
166. List the insulating materials commonly used in electrical work at the present time.
167. List the conducting materials commonly used in electrical work at the present time.
168. What is the purpose of the sheath of a cable?
169. Compare a married straight joint and a straight weak-back ferrule joint.
170. What is the maximum resistance of a cable joint?
171. Why should acidic soldering fluxes never be used in cable joints?
172. Why is a nicked conductor likely to cause trouble?
173. What is meant by the 'ambient air temperature'?
174. What is the difference between a 'live' conductor and a 'dead' conductor?
175. How should a fllexible cord be extended?
176. Why are bell contact points made of German silver or platinum?
177. In what circumstances may a trunking be used to contain cables insulated for low voltage and cables insulated for extra-low voltage?
178. In what circumstances is cleated wiring used?
179. Can the neutral conductor be looped from another final subcircuit?
180. State four possible causes of corrosion of metal sheaths of cables in damp situations.
181. Define a ring circuit.

182. What are the IEE requirements regarding the installation of an earthed-concentric wiring system?
183. How does a thermostat maintain temperature within limits?
184. What is the National Grid?
185. Compare Class A and Class B conduits as wiring systems.
186. What is meant by 'bonding'?
187. What voltage should be used in an IR test? Why?
188. Describe briefly the need for carrying out a visual inspection on an installation.
189. Why are additional precautions necessary in a room containing a bath?
190. In what circumstances should twin-twisted flex not be used for a lighting pendant?
191. What is meant by a 'flameproof' fitting?
192. Describe the sequence of supply-control equipment at a supply-intake position.
193. What is meant by a 4-gang switch?
194. Describe an 'all-insulated' wiring system and its uses.
195. Describe uses for the double-pole switch.
196. What is the reason for differences in the BC and ES lampholder?
197. What is the effect of voltage drop in a circuit?
198. What is meant by a 'leakage current'?
199. Why is a limit placed on the number of conductors that can be drawn into a conduit?
200. Obtain an electrical accessory made, say, thirty years ago. Compare it in all respects with the same accessory as used today.

List of IEE Regulations mentioned in the text

Index